Nehemiah Hawkins

Maxims and Instructions for the Boiler Room

Nehemiah Hawkins

Maxims and Instructions for the Boiler Room

ISBN/EAN: 9783744758420

Printed in Europe, USA, Canada, Australia, Japan

Cover: Foto ©berggeist007 / pixelio.de

MAXIMS

AND

INSTRUCTIONS

FOR

THE BOILER ROOM.

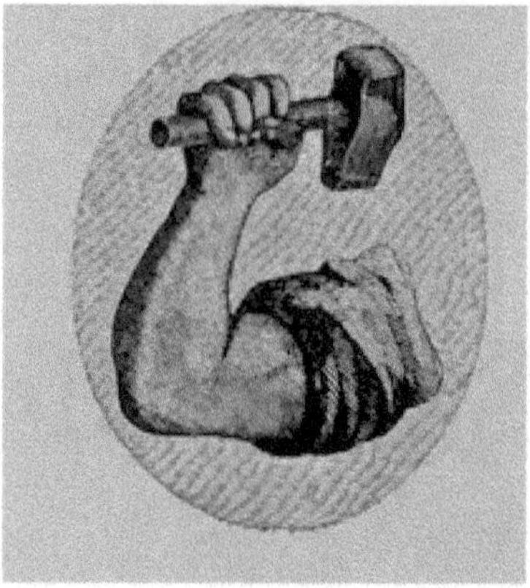

This Work is Fraternally inscribed to W. R. Hawkins, R. F. Hawkins and F. P. Hawkins.

RICHARD TREVETHICK

MAXIMS AND INSTRUCTIONS

FOR

The Boiler Room.

USEFUL TO

Engineers, Firemen & Mechanics,

RELATING TO STEAM GENERATORS, PUMPS APPLIANCES, STEAM HEATING, PRACTICAL PLUMBING, ETC.

BY N. HAWKINS, M. E.,

HONORARY MEMBER NATIONAL ASSOCIATION OF STATIONARY ENGINEERS, EDITORIAL WRITER, AUTHOR OF HAND BOOK OF CALCULATIONS FOR ENGINEERS AND FIREMEN, ETC., ETC.

COMPRISING INSTRUCTIONS AND SUGGESTIONS ON THE CONSTRUCTION, SETTING, CONTROL AND MANAGEMENT OF VARIOUS FORMS OF STEAM BOILERS; ON THE THEORY AND PRACTICAL OPERATION OF THE STEAM PUMP; STEAM HEATING; PRACTICAL PLUMBING; ALSO RULES FOR THE SAFETY VALVE, STRENGTH OF BOILERS, CAPACITY OF PUMPS, ETC.

THEO. AUDEL & CO., Publishers.

63 FIFTH AVE., COR. 13TH ST.,

NEW YORK.

1898.

PREFACE.

The chief apology for the preparation and issue of these Maxims and Instructions, for the use of Steam users, Engineers and Firemen, is the more than kind reception of Calculations.

But there are other reasons. There is the wholesome desire to benefit the class, with whom, in one way and another, the author has been associated nearly two score years.

The plan followed in this work will be the same as that so generally approved in Calculations; the completed volume will be a work of reference and instruction upon those works set forth in the title page. As a work of reference the work will be especially helpful through combined Index and Definition Tables to be inserted at the close of the book. By the use of these the meaning of every machine, material and performance of the boiler room can be easily found and the "points" of instruction made use of.

This work being issued in parts, now in manuscript, and capable of change or enlargement, the editor will be thankful for helpful suggestions from his professional brethren, before it is put into permanent book form.

N. HAWKINS.

OLIVER EVANS
GEORGE STEPHENSON
ROBERT FULTON

INTRODUCTION.

Each successive generation of engineers has added certain *unwritten experiences* to the general stock of knowledge relating to steam production, which have been communicated to their successors, and by them added to, in their turn, it is within the province of this book to put in form for reference, these unwritten laws of conduct, which have passed into MAXIMS among engineers and firemen—a maxim being an undisputed truth, expressed in the shortest terms.

SOLILOQUY OF AN ENGINEER. "Standing in the boiler room and looking around me, there are many things I ought to know a good deal about. Coal! What is its quality? How much is used in ten hours or twenty-four hours? Is the grate under the boiler the best for an economical consumption of fuel? Can I, by a change in method of firing save any coal? The safety valve. Do I know at what pressure it will blow off? Can I calculate the safety-valve so as to be certain the weight is placed right? Do I know how to calculate the area of the grate, the heating surface of the tubes and shell? Do I know the construction of the steam gauge and vacuum-gauge? Am I certain the steam-gauge is indicating correctly, neither over nor under the pressure of the steam? What do I know about the setting of boilers? About the size and quality of fire bricks? About the combination of carbon and hydrogen of the fuel with the oxygen of the atmosphere? About oxygen, hydrogen and nitrogen? About the laws of combustion? About radiation and heating surfaces?

"Do I know what are good non-conductors for covering of pipes, and why they are good? Do I know how many gallons of water are in the boiler?

"What do I know about water and steam? How many gallons of water are evaporated in twenty-four hours? What do I know about iron and steel, boiler evaporation, horse-power of engines, boiler appendages and fittings?

"Can I calculate the area and capacity of the engine cylinder? Can I take an indicator diagram and read it? Can I set the eccentric? Can I set valves? Do I understand the construction of the thermometer, and know something about the pressure of the atmosphere, temperature and the best means for ventilation? Can I use a pyrometer and a salinometer?

"Without going outside of my boiler and engine room I find these things are all about me—air, water, steam, heat, gases, motion, speed, strokes and revolutions, areas and capacities—how much do I know about these?

"How much can be learned from one lump of coal? What was it, where did it come from? When it is burned, what gases will it give off?

"And so with water. What is the composition of water? What are the effects of heat upon it? How does it circulate? What is the temperature of boiling water? What are the temperatures under different pressures? What is latent heat? What is expansive force?"

These are the questioning thoughts which fill, while on duty, more or less vividly, the minds of both engineers and firemen, and it is the purpose of this volume to answer the enquiries, as far as may be without attempting too much; for the perfect knowledge of the operations carried on within the boiler-room involves an acquaintance with many branches of science. In matters relating to steam engineering, it must be remembered that "art is long and time is short."

The utility of such a book as this is intended to be, no one will question, and he who would not be a "hewer of wood and a drawer of water" to the more intelligent and well-informed mechanic, must possesses to a considerable extent the principles and rules embraced in this book; and more especially, if he would be master of his profession and reputed as one whose skill and decisions can be implicitly relied upon.

The author in the preparation of the work has had two objects constantly in view; first to cause the student to become familiarly acquainted with the leading principles of his profession as they are mentioned, and secondly, to furnish him with as much advice and information as possible within the reasonable limits of the work.

While it is a fact that some of the matter contained in this work is very simple, and all of it intended to be very plain, it yet remains true that the most expert living engineer was at one time ignorant of the least of the facts and principles here given, and at no time in his active career can he ever get beyond the necessity of knowing the primary steps by which he first achieved his success.

The following taken from the editorial columns of the leading mechanical journal of the country contain the same suggestive ideas already indicated in the "soliloquy of an engineer:"

"There is amongst engineers in this country a quiet educational movement going on in matters relating to facts and principles underlying their work that is likely to have an important influence on industrial affairs. This educational movement is noticeable in all classes of workmen, but amongst none more than among the men in charge of the power plants of the country. It is fortunate that this is so, for progress once begun in such matters is never likely to stop.

"Engineers comprise various grades from the chief engineer of some large establishment, who is usually an accomplished mechanic, carrying along grave responsibilities, to the mere stopper and starter, who is engineer by courtesy only, and whose place is likely to be soon filled by quite another man, so far as qualifications are concerned. Men ignorant of everything except the mere mechanical details of their work will soon have no place.

"Scarcely a week passes that several questions are not asked by engineers, either of which could be made the subject of a lengthy article. This is of interest in that it shows that engineers, are not at the present time behind in the way of seeking information. Out of about a thousand questions that went into print, considerable more than half were asked by stationary engineers. These questions embrace many things in the way of steam engineering, steam engine management, construction, etc."

The old meaning of the word lever was "a lifter" and this book is intended to be to its attentive student, a real lever to advance him in his life work; it is also to be used like a ladder, which is to be ascended step by step, the lower rounds of which, are as important as the highest.

It is moreover, the earnest wish of the editor that when some, perchance may have "climbed up" by the means of this, his work, they may in their turn serve as lifters to advance others, and by that means the benefits of the work widely extended.

MATERIALS.

The things with which the engineer has to deal in that place where steam is to be produced as an industrial agent, are

1. The Steam Generator.

2. Air.

3. Fuel.

4. Water.

5. Steam Appliances.

Starting with these points which form a part of every steam plant, however limited, however vast, the subject can easily be enlarged until it embraces a thousand varied divisions extending through all time and into every portion of the civilized world.

It is within the scope of this work to so present the subjects specified, that the student may classify and arrange the matter into truly scientific order.

MATERIALS.

In entering the steam department, where he is to be employed, the eye of the beginner is greeted with the sight of coal, water, oil, etc., and he is told of invisible materials, such as air, steam and gases; it is the proper manipulation of these seen and unseen material products as well as the machines, that is to become his life task. In aiding to the proper accomplishment of the yet untried problems nothing can be more useful than to know something of the nature and history of the different forms of matter entering into the business of steam production. Let us begin with

COAL.

The source of all the power in the steam engine is stored up in coal in the form of heat.

And this heat becomes effective by burning it, that is, by its combustion.

Coal consists of carbon, hydrogen, nitrogen, sulphur, oxygen and ash. These elements exist in all coals but in varying quantities.

These are the common proportions of the best sorts:

	ANTHRACITE	BITUMINOUS	WOOD (AVERAGE) DRY.	PEAT	PEAT ¼ WATER
Carbon ...	90½	81	50	59	44
Hydrogen .	2½	5¼	6	6	4½
Nitrogen ..	0¼	1	1	1¼	1
Sulphur ...	00	1½	0	?	(25)
Oxygen. ..	2½	6½	41	30	22½
Ash......	4¼	4¾	2	3¾	3
	100	100	100	100	100

In burning coal or other fuel atmospheric air must be introduced before it will burn; the air furnishes the oxygen, without which combustion cannot take place.

MATERIALS.

It is found that in burning one lb. of coal one hundred and fifty cubic feet of air must be used and in every day practice it is necessary to supply twice as much ; this is supplied to the coal partly through the grate bars, partly through the perforated doors, and the different devices for applying it already heated to the furnace.

WOOD.

Wood as a combustible, is divisible into two classes : 1st, the hard, compact and comparatively heavy, such as oak, ash, beech, elm. 2d, the light colored soft, and comparatively light woods as pine, birch, poplar.

Wood when cut down contains nearly half moisture and when kept in a dry place, for several years even, retains from 15 to 20 per cent. of it.

The steam producing power of wood by tests has been found to be but little over half that of coal *and the more water in it the less its heating power*. In order to obtain the most heating power from wood it is the practice in some works in Europe where fuel is costly, to dry the wood fuel thoroughly, even using stoves for the purpose, before using it. This "hint" may serve a good purpose on occasion.

The composition of wood reduced to its elementary condition will be found in the table with coal.

PEAT.

Peat is the organic matter or vegetable soil of bogs, swamps and marshes—decayed mosses, coarse grasses, etc. The peat next the surface, less advanced in decomposition, is light, spongy and fibrous, of a yellow or light reddish-brown color ; lower down it is more compact, of a darker-brown color, and in the lowest strata it is of a blackish brown, or almost a black color, of a pitchy or unctuous feel.

Peat in its natural condition generally contains from 75 to 80 per cent. of water. It sometimes amounts to 85 or 90 per cent. in which case the peat is of the consistency of mire.

When wet peat is milled or ground so that the fibre is broken, crushed or cut, the contraction in drying is much increased by this treatment; and the peat becomes denser, and is better consolidated than when it is dried as it is cut from the bog; peat so prepared is known as *condensed* peat, and the degree of condensation varies according to the natural heaviness of the peat. So effectively is peat consolidated and condensed by the simple process of breaking the fibres whilst wet, that no merely mechanical force of compression is equal to it.

In the table the elements of peat are presented in two conditions. One perfectly dried into a powder before analyzing and the other with 25 per cent. of moisture.

The value of peat as a fuel of the future is an interesting problem in view of the numerous inroads made upon our great natural coal fields.

TAN.

Tan, or oak bark, after having been used in the process of tanning is burned as fuel. The spent tan consists of the fibrous portion of the bark. Five parts of oak bark produce four parts of dry tan.

STRAW.

Two compositions of straw (as a fuel) is as follows:

Water, - - - -	14	per cent.
Combustible matter, - -	79	"
Ash, - - - -	7	"

COKE, CHARCOAL, PEAT CHARCOAL.

These are similar substances produced by like processes from coal, wood, and peat and they vary in their steam-producing power according to the value of the fuels from which they are produced. The method by which they are made is termed carbonization, which means that all the gases are removed by heat in closed vessels or heaps, leaving only the carbon and the more solid parts like ashes.

LIQUID AND GAS FUELS.

Under this head come petroleum and coal gas, which are obtained in great variety and varying value from coal and coal oil. The heating power of these fuels stands in the front rank, as will be seen by the table annexed.

MATERIALS.

There are kinds of fuel other than coal such as wood, coke, sawdust, tan bark, peat and petroleum oil and the refuse from oil. These are all burned with atmospheric air of which the oxygen *combines* with the combustible part of the fuel while the nitrogen passes off into the chimney as waste.

The combustible parts of coal are carbon, hydrogen and sulphur and the unburnable parts are nitrogen, water and the incombustible solid matters such as ashes and cinder. In the operation of firing under a boiler the three first elements are totally consumed and form heat ; the nitrogen, and water in the form of steam, escapes to the flue, and the ashes and cinders fall under the grates.

The anthracite coal retain their shape while burning, though if too rapidly heated they fall to pieces. The flame is generally short, of a blue color. The coal is ignited with difficulty ; it yields an intense local or concentrated heat ; and the combustion generally becomes extinct while yet a considerable quantity of the fuel remains on the grate.

The dry or free burning bituminous coals are rather lighter than the anthracites, and they soon and easily arrive at the burning temperature. They swell considerably in coking, and thus is facilitated the access of air and the rapid and complete combustion of their fixed carbon.

The method of firing with different sorts of fuel will be treated elsewhere.

AIR.

The engineer's success in the management of the furnace depends quite as much upon his handling the air in the right mixtures and proportions as it does in his using the fuel—for

1. Although invisible to the eye air is as much *a material substance* as coal or stone. If there were an opening into the interior of the earth which would permit the air to descend its density would increase in the same manner as it diminishes in the opposite direction. At the depth of about 34 miles it would be as dense as water, and at the depth of 48 miles it would be as dense as quicksilver ; and at the depth of about 50 miles as dense as gold.

MATERIALS.

2. Air is not only a substance, but *an impenetrable body ;* as for example : if we make a hollow cylinder, smooth and closed at the bottom, and put a stopper or solid piston to it, no force will enable us to bring it into contact with the bottom of the cylinder, unless we permit the air within it to escape.

3. Air is *a fluid* which is proved by the great movability of its parts, flowing in all directions in great hurricanes and in gentle breezes ; and also by the fact that a pressure or blow is propagated through all parts and affects all parts alike.

4. It is also an *elastic fluid,* thus when an inflated bladder is compressed it immediately restores itself to its former situation; indeed, since air when compressed restores itself or tends to restore itself, with the same force as that with which it is compressed, it is a perfectly elastic body.

5. The weight of a column of air one square foot at the bottom is found to be 2156 lbs. or very nearly 15 lbs. to the square inch, hence it is common to state the pressure of the atmosphere as equal to 15 lbs. to the square inch.

It follows from these five points that the engineer must consider air as a positive, although unseen, factor with which his work is to be accomplished.

What air is composed of is a very important item of knowledge. It is made of *a mixture* of two invisible gases whose minute and inconceivably small atoms are mingled together like a parcel of marbles and bullets—that is while together they do not lose any of their distinctive qualities. The two gases are called nitrogen and oxygen, and of 100 parts or volumes of air 79 parts are of nitrogen and 21 parts of oxygen ; but *by weight* (for the oxygen is the heaviest) 77 of nitrogen and 23 of oxygen.

The oxygen is the part that furnishes the heat by uniting with the coal—indeed without it the process of combustion would be impossible : of the two gases the oxygen is burned in the furnace, more or less imperfectly, and the nitrogen is wasted.

MATERIALS.

TABLE OF EVAPORATION.

In order to arrive at the money value of the various fuels heretofore described a method of composition has been arrived at which gives very accurately their comparative worth. The rule is too advanced for this elementary work, but the following results are plainly to be understood, and will be found to be of value.

Lbs. of Fuel.	Temperature of water 212°
Coal,	14.62 lbs. of water.
Coke,	14.02 "
Wood,	8.07 "
Wood, 25 % of water,	6.05 "
Wood Charcoal,	13.13 "
Peat, perfectly dry,	10.30 "
Peat with 25 % moisture,	7.41 "
Peat, Charcoal (dry)	12.76 "
Tan, dry,	6.31 "
Tan, 30 % moisture,	4.44 "
Petroleum,	20.33 "
Coal gas 1 lb. or (31½ cut feet)	47.51 "

The way to read this table is as follows: "One lb. coal has an average evaporative capacity of $14.\frac{62}{100}$ lbs. of water," or

One lb. of peat with one-quarter moisture will evaporate, if *all* the heat is utilized $7.\frac{41}{100}$ lbs. of water.

In practice but little over half of these results are attained, but for a matter of comparison of the value of one kind of fuel with another the figures are of great value; a boiler burning wood or tan needs to be much larger than one burning petroleum oil.

FIRE IRONS.

The making or production of steam requires the handling of the fuel, more or less, until its destruction is complete, leaving nothing behind in the boiler room, except ashes and clinkers. The principal tools used by the attendant, to do the task most efficiently, are: 1. The scoop shovel. 2. The poker. 3. The slice bar. 4. The barrow.

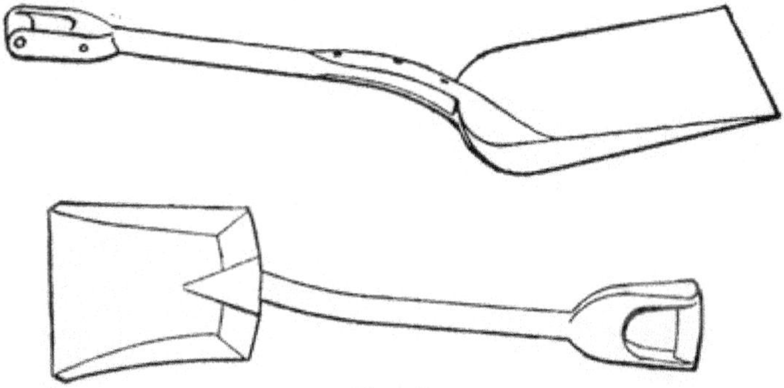

Fig. 1.

Fig. 1. represents the regular scoop shovel commonly called "a coal shovel," but among railroad men and others, known as a locomotive or charging scoop. The cut also represents a regular shovel. Both these are necessary for the ordinanry business of the boilor room.

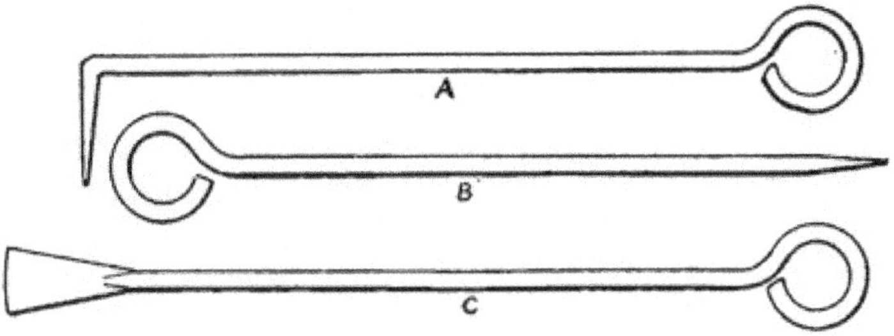

Fig. 2.

In cut 2 are represented a furnace poker, A, and two forms of the slice bar. They are all made by blacksmiths from round iron, some 7 or 8 feet long and only vary in the form of the end. The regular slice bar is shown in C, Fig. 2; and "the dart" a special form used largely on locomotives is shown in B.

FIRE IRONS.

The dexterous use of these important implements can merely be indicated in print, as it is part of the trade which is imparted by oral instruction. One "point" in making the slice bar may be mentioned to advantage—the lower side should be perfectly flat *so that it may slide* on the surface of the grate bars as it is forced beneath the fire—and the upper portion of the edge should be in the shape of a half wedge, so as to crowd upwards the ashes and clinkers while the lower portion slides along.

There is sometimes used in connection with these tools an appliance called a LAZY BAR. This is very useful for the fireman when cleaning a bituminous or other coal fire: it saves both time and fuel as well as steam. It is a hook shaped iron, ingeniously attached above the furnace door, so that it supports the principal part of the weight of the heavy slice bar or poker when being used in cleaning out the fires

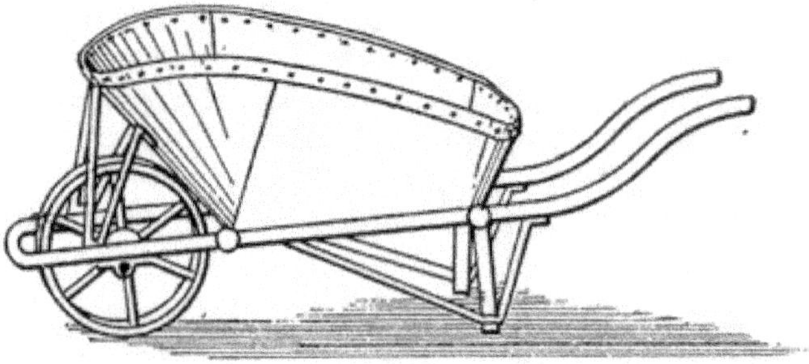

Fig. 3.

Equally necessary to the work of the boiler-room is the barrow shown in cut. There are many styles of the vehicle denominated respectively—the railroad barrow, the ore and stone barrow, the dirt barrow, etc.; but the one represented in fig. 3 is the regular coal barrow.

In conveying coal to "batteries" of boilers, in gas houses and other suitable situations the portable car and iron track are nearly always used instead of the barrow. In feeding furnaces with saw dust and shavings large iron screw conveyors are frequently employed, as well as blowers—In the handling of the immense quantities of fuel required, the real ingenuity of the engineer in charge has ample opportunity for exercise.

FIRE IRONS.

There are also used in nearly all boiler rooms HOES made of heavy plate iron, with handles similar to those shown in the cuts representing the slice bar and poker. A set of two to four hoes of various sizes is a very convenient addition to the list of fire tools; a light garden hoe for handling ashes is not to be omitted as a labor saving tool.

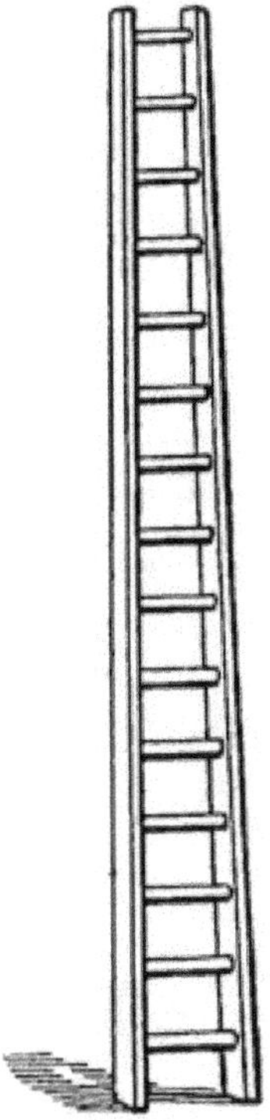

Fig. 4.

HANDY TOOLS.

Besides the foregoing devices for conducting the preliminary process of the steam generation, the attendant should have close at hand a servicable HAND HAMMER, a SLEDGE for breaking coal and similar work, and A SCREW WRENCH and also a light LADDER for use about the boiler and shafting.

In addition to these there are various other things almost essential for the proper doing of the work of the boiler room,—FIRE AND WATER PAILS, LANTERNS, RUBBER HOSE, etc., which every wise steam user will provide of the best quality and which the engineer will as carefully keep in their appointed places ready for instant service.

To these familiar tools can be added FILES, LACE CUTTERS, BOILER-FLUE BRUSHES, STOCK and DIES, PIPE-TONGS, SCREW JACKS, VISES, etc., all of which when used with skill and upon right occasion pay a large return on their cost.

THE TOOL BOX.

The complex operations of the boiler room, its emergencies and varying conditions demand the use of many implements which might at first thought be out of place. The following illustrations exhibit some of these conveniences.

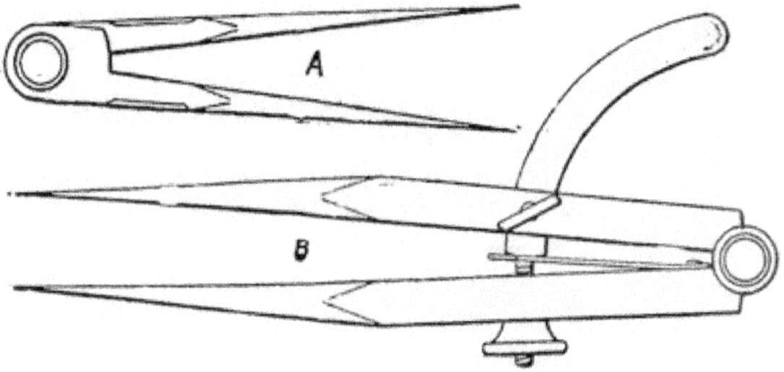

Fig. 5.

Fig. 5, letter A, show the common form of COMPASSES which are made from 3 to 8 inches long. Letter B, illustrates the common steel compass dividers, which are made from 5 to 24 inches in length.

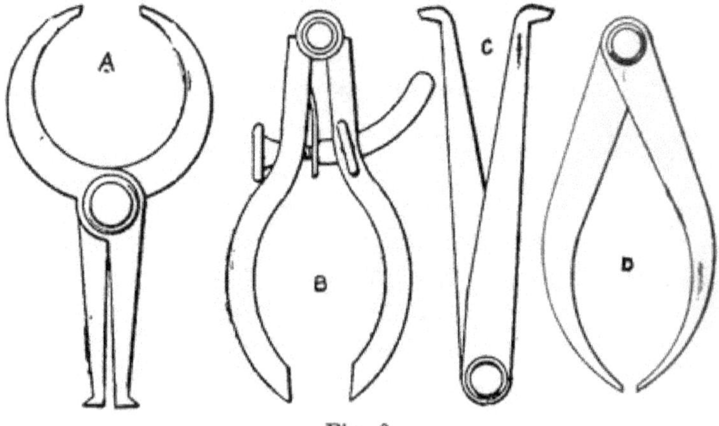

Fig. 6.

In this illustration, A exhibits double, inside and outside CALIPERS; B, adjustable outside Calipers; C, inside; and D outside, plain caliapers.

FIRING OF

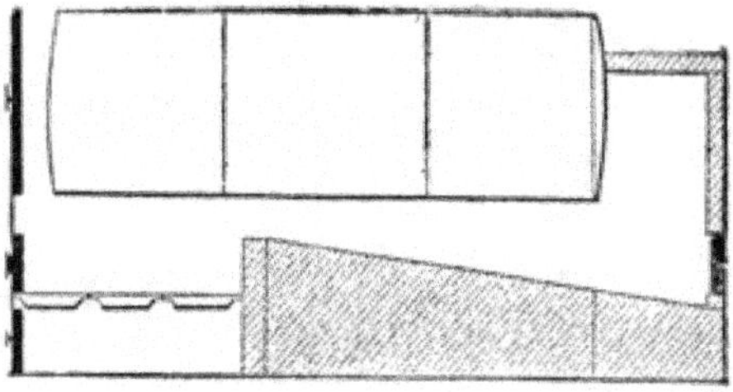

STEAM BOILERS.

THE FIRING OF STEAM BOILERS.

The care and management of a steam boiler comprises three things:

1. The preparation, which includes the partial filling with water and the kindling of the fire.

2. The running, embracing the feeding, firing and extinction or banking of the fire.

3. The cleaning out after it has been worked for some time.

To do this to the best advantage, alike to owner and employee, can be learned only by practice under the tuition of an experienced person. The "trick" or unwritten science of the duties of the skillful firemen must be communicated to the beginner, by already experienced engineers or firemen or from experts who have made the matter a special study. *Let it be understood that the art of firing cannot be self taught.*

The importance of this knowledge is illustrated by a remarkable difference shown in competitive tests in Germany between trained and untrained firemen in the matter of securing a high evaporation per pound of coal. The trained men succeeded in evaporating 11 lbs. of water, as against 6.89 lbs. which was the best that the untrained men could obtain.

It is certain that a poor fireman is a dear man at any price, and that a competent one may be cheap at twice the wages now paid. Suppose, for instance, a man who burns three tons a day is paid $2.00 for such service, and that in so doing he is wasting as little as 10 per cent. If the coal cost $4.50 per ton the loss will be $1.35 per day, or what is equivalent to paying a man $3.35 per day who can save this amount.

The late Chief Engineer of Philadelphia Water Works effected an annual saving to the city of something like $50,000; and recently the weekly consumption of a well established woolen mill was reduced from 71 to 49 tons, a clear saving of 22 tons by careful attention to this point.

THE FIRING OF STEAM BOILERS.

It is apparent that any rules or directions which might be given for one system would not apply equally to other forms of boilers and this may be the principal reason that the art is one so largely of personal instruction. Some rules and hints will, however be given to the beginner, which may prove of advantage in fitting the fireman for an advanced position; or to assure him permanence in his present one.

No two boilers alike. It is said that no two boilers, even though they seemed to be exactly alike—absolute duplicates—ever did the same, or equal service. Every steam boiler, like every steam engine, has an individuality of its own, with which the person in charge has to become acquainted, in order to obtain the best results from it.

The unlikeness in the required care of steam engines which seem to be exactly the same, is still more marked in the different skill and experience demanded in handling locomotive, marine, stationery, portable boilers and other forms of steam generators.

BEFORE LIGHTING THE FIRE under the boiler in the morning, the engineer or fireman should make a rapid yet diligent examination of various things, viz.: 1. He should make sure that the boiler has the right quantity of water in it—that it has not run out during the night or been tampered with by some outside party; very many boilers have been ruined by neglecting this first simple precaution. 2. He should see that the safety valve is in working order; this is done by lifting by rod or hand the valve which holds the weight upon the safety valve rod. 3. He should open the upper gauge-cock to let out the air from the boiler while the steam is forming. 4. He should examine the condition of the grate-bars and see that no clinkers and but few ashes are left from last night's firing. 5. And finally, after seeing that everything is in good shape, proceed to build the fire as follows:

ON LIGHTING THE FIRE. When quite certain that everything is in good condition, put a good armful of shavings or fine wood upon the grate, then upon this some larger pieces of wood to form a bed of coals, and then a little of the fuel that is to be used while running. Sometimes it is better to light before putting on the regular fuel, but in any case give it plenty of air. Close the fire doors, and open the ash pit, giving the chimney full draught.

When the fire is well ignited, throw in some of the regular fuel, and when this is burning add more, a little at a time, and continue until the fire is in its normal condition, taking care, however, not to let it burn too freely for fear of injury to the sheets by a too rapid heating.

It is usually more convenient to light the fire through the fire door, but where this cannot be done, a torch may be used beneath the grates, or even a light fire of shavings may be kindled in the ash pit.

At the time of lighting, all the draughts should be wide open.

As soon as the steam is *seen* to issue from the open upper gauge-cock it is proof that the air is out. It should now be closed and the steam gauge will soon indicate a rise in temperature.

When the steam begins to rise it should next be observed that: 1. All the cocks and valves are in working order—that they move easily. 2. That all the joints and packings are tight.

In the following two cuts are exhibited in an impressive way the difference between proper and improper firing.

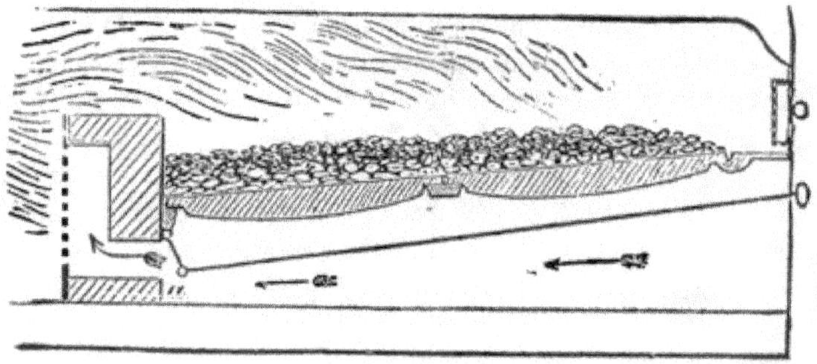

Fig. 1.

Fig. 1 represents the proper mode of keeping an even depth of coal on the grate bars; the result of which will be, a uniform generation of gas throughout the charge, and a uniform temperature in the flues.

THE FIRING OF STEAM BOILERS.

Fig. 2.

Fig. 2 represents a very frequent method of feeding furnaces: charging the front half as high, and as near the door, as possible, leaving the bridge end comparatively bare. The result necessarily is that more air obtains access through the uncovered bars than is required, which causes imperfect combustion and consequent waste.

The duties of the fireman in the routine of the day may thus be summed up:

1st.—*Begin to charge the furnace at the bridge end and keep firing to within a few inches of the dead plate.*

2d.—*Never allow the fire to be so low before a fresh charge is thrown in, that there shall not be at least three to five inches deep of clean, incondescent fuel on the bars, and equally spread over the whole.*

3d.—*Keep the bars constantly and equally covered,* particularly at the sides and the bridge end, where the fuel burns away most rapidly.

4th.—If the fuel burns unequally or into holes, *it must be leveled, and the vacant spaces must be filled.*

5th.—The large coals must be broken into pieces not bigger than a man's fist.

6th.—When the ash pit is shallow, it must be the more frequently cleared out. A body of hot cinders, beneath them, overheats and burns the bars.

7th.—The fire must not be hurried too much, but should be left to increase in intensity gradually. When fired properly the fuel is consumed in the best possible way, no more being burned than is needed for producing a sufficient quantity of steam and keeping the steam pressure even.

DIRECTIONS FOR FIRING WITH VARIOUS FUELS.

FIRING BOILERS NEWLY SET, etc.—Boilers newly set should be heated up very slowly indeed, and the fires should not be lighted under the boilers for at least two weeks after setting, if it is possible to wait this length of time. This two weeks enables all parts of the mason work to set gradually and harden naturally; the walls will be much more likely to remain perfect, than when fires are lighted while the mortar is yet green.

When fire is started under a new boiler the first time, it should be a very small one, and no attempt should be made to do more than moderately warn. all parts of the brick work. A slow fire should be kept up for twenty-four hours, and on the second day it may be slightly increased. Three full days should elapse before the boiler is allowed to make any steam at all.

When the pressure rises, it should not be allowed to go above four or five pounds and the safety valve weight should be taken off to prevent any possibility of an increase. Steam should be allowed to go through all the pipes attached for steam, and blow through the engine before any attempt is made to get pressure on them. The object of all these precautions and this care is to prevent injury by sudden expansion, which, may cause great damage.

FIRING WITH COKE.

Coke, in order to be completely consumed, needs a greater volume of air per pound of fuel than coal. Theoretically it needs from 9 to 10 lbs. of air to burn a pound of coal, and 12 to 13 lbs. of air to burn a pound of coke.

Coke, therefore, requires a more energetic draft, which is increased by the fact that it can only burn economically in a thick bed. It is also necessary to take into account the size of the pieces.

The ratio between the heating and grate surface should be less with coke than with coal, that is to say, the grate should be larger.

The difference amounts to about 33 per cent. In fact,

FIRING WITH VARIOUS FUELS.

about $9\frac{3}{4}$ lbs. of coke should be burned per hour on each square foot of grate area, while at least $14\frac{1}{2}$ lbs. of coal can be burned upon the same space.

The high initial temperature which is developed by the combustion of coke requires conducting walls. Therefore the furnace should not be entirely surrounded by masonry; and the plates of the boiler should form at least the crown of the fire-box. In externally fired boilers, the furnace should be located beneath and not in front of the boiler. Internal fire boxes may be used, but the greatest care should be exercised to avoid any incrustation of the plates, and in order that this may be done, only the simplest forms of boilers should be used. With coke it is not essential that long passages should be provided for the passage of the products of combustion, since the greater part of the heat developed is transmitted to the sheets in the neighborhood of the furnace.

Since coke contains very little hydrogen, the quick flaming combustion which characterizes coal is not produced, but the fire is more even and regular. And, finally, the combustion of coal is distinguished by the fact that in the earlier phases there is usually an insufficiency of air, while in the last there is no excess.

The advantage of coke over raw soft coal as a fuel is that otherwise useless slack can be made available by admixture in its manufacture, and especially that it can be perfectly and smokelessly burnt without the need of skilled labor. And we cannot doubt that the public demand for a clear and healthy atmosphere will finally result in the almost complete substitution of coke fuel for soft lump coal.

SIXTEEN STEAM BOILERS in a large mill in Massachusetts of 54 and 60 inches in diameter are fired as follows:

There are three separate batteries; one of five boilers, one of twelve and one of three. Each boiler is fired every five minutes. There are two firemen for the battery of twelve and

FIRING WITH VARIOUS FUELS.

one for each of the others. A gong in each fire-room is operated by electricity in connection with a clock. The duty of the fireman is this, that when the gong strikes he commences at one end of his fire-room and fires as rapidly as possible, opening one-half of each furnace door. The coal is thrown only on one-half of the grate space as he rapidly fires each boiler, the other half is covered at the next sounding of the gong. The old style of straight grate is used. The fires are kept six inches thick or a little thicker. No slicing is done. It is, of course, to be understood that the firemen arrange the quantity of coal fired according to the apparent necessity of the case. Bituminous coal is used, and it is broken into small pieces so as to distribute well. Accurate account is kept of the quantity of coal used and the engines are frequently indicated.

TWENTY HORSE POWER.—An old engineer says the way he handled his boiler of this size, burning 800 lbs. of screenings per day, is as follows:

My method is to run as heavy a fire as my fire box will allow to be kept under the bridge wall, and not to disturb it more than once in a ten hours run, then clean out with care and as speedily as possible, dress light and let it come up and get ready to bank. In banking I make sure to have an even fire, as deep as the bridge wall will allow. Then I shut my dampers and let it lie. In the morning I open and govern by the dampers. I do not touch my fire until 3.30 or 4 o'clock in the afternoon, and then proceed to clean as before.

FIRING WITH COAL TAR.—The question of firing retort benches with tar instead of coke has engaged the attention of gas managers for many years, and various modes have been adopted for its management. The chief difficulty has been in getting a constant flow of tar into the furnace, uninterrupted by stoppages caused by the regulating cock or other appliance not answering its purpose and by the carbonizing of the tar in the delivery pipe, thus choking it up and rendering it uncertain in action. To obviate these

FIRING WITH VARIOUS FUELS.

difficulties various plans have been resorted to, but the best means for overcoming them are thus described ; fix the tar supply tank as near the furnace to be supplied as convenient, and one foot higher than the tar-injector inlet A cock is screwed into the side of the tank, to which is attached a piece of composition pipe ⅜-inch in diameter, ten inches long. To this a ½-inch iron service pipe is connected, the other end of which is joined to the injector. By these means it is found that at the ordinary temperature of the tar well (cold weather excepted) four gallons of tar per hour are delivered in a constant steam into the furnace. If more tar is required, the piece of ⅜-inch tube must be shortened, or a larger tube substituted, and if less tar is required it must be lengthened. The risk of stoppage in the nozzle of the injector is overcome by the steam jet, which scatters the tar into spray and thus keeps everything clear. Trouble being occasioned by the retorts becoming too hot, in which case, on shutting off the flow of tar for a while, the tar in the pipe carbonized and caused a stoppage, a removable plug injector is fitted and ground in like the plug of a cock, having inlets on either side for tar and steam. This plug injector can be removed, the tar stopped in two seconds and refixed in a similar time. The shell of the injector is firmly bolted to the top part of the door frame. The door is swung horizontally, having a rack in the form of a quadrant, by which it is regulated to any required height, and to admit any quantity of air.

FIRING WITH STRAW.—The operation of burning straw under a boiler consists in the fuel being fed into the furnace only as fast as needed. When the straw is handled right, it makes a beautiful and very hot flame and no smoke is seen coming from the stack. The whole secret of getting the best results from this fuel is to feed it into the furnace in a gradual stream as fast as consumed. When this is done complete combustion is the result. A little hole may be drilled in the smoke-box door, so that the color of the fire can be seen and fire is handled accordingly. When the smoke comes from the stack the color of the flame is that of a good gas jet. By feeding a

FIRING WITH VARIOUS FUELS.

little faster the color becomes darker and a little smoke comes from the stack; feeding a little faster the flame gets quite dark and the smoke blacker; faster still, the flame is extinguished, clouds of black smoke come from the stack, and the pressure is falling rapidly.

FIRING WITH OIL.—Great interest is now manifested in the use of oil as fuel. There are various devices used for this purpose, most of them depending upon a steam jet to atomize the oil, or a system of retorts to first heat the oil and convert it into gas, before being burned.

Another method in successful operation is the use of compressed air for atomizing the oil—air being the element, nature provides for the complete combustion of all matter. The cleanliness of the latter system and its comparative freedom from any odor of oil or gas and its perfect combustion, all recommend it. Among the advantages claimed for the use of oil over coal are 1, uniform heat; 2, constant pressure of steam; 3, no ashes, clinkers, soot or smoke, and consequently clean flues; 4, uniform distribution of heat and therefore less strain upon the plates.

FIRING ON AN OCEAN STEAMER like the "*Umbria*," the men come on in gangs of eighteen stokers or firemen and twelve coal passers, and the "watch" lasts four hours. The "*Umbria*" has 72 furnaces, which require nearly 350 tons of coal a day, at a cost of almost $20,000 per voyage. One hundred and four men are employed to man the furnaces, and they have enough to do. They include the chief engineer, his three assistants, and ninety stokers and coal passers.

The stoker comes to work wearing only a thin undershirt, light trousers, and wooden shoes. On the "*Umbria*" each stoker tends four furnaces. He first rakes open the furnaces, tosses in the coal, and then cleans the fire, that is, pries the coal apart with a heavy iron bar, in order that the fire may burn freely. He rushes from one furnace to another, spending perhaps two or three minutes at each. Then he dashes to the air pipe, takes his turn at cooling off, and waits for another

FIRING WITH VARIOUS FUELS.

call to his furnace, which comes speedily. When the "watch" is over, the men shuffle off, dripping with sweat from head to foot, through long, cold galleries to the forecastle, where they turn in for eight hours. Four hours of scorching and eight hours sleep make up the routine of a fireman's life on a voyage.

The temperature is ordinarily 120°, but sometimes reaches 160°; and the work then is terribly hard. The space between the furnaces is so narrow that when the men throw in coal they must take care when they swing back their shovels, lest they throw their arms on the furnace back of them.

In a recent trial of a government steamer the men worked willingly in a temperature of 175°, which, however, rose to 212° or the heat of boiling water. The shifts of four hours were reduced to two hours each, but after sixteen men had been prostrated, the whole force of thirty-six men refused to submit to the heat any longer and the trial was abandoned.

There is no place on ocean or land, where more suffering is inflicted and endured by human beings than in these h—— holes, quite properly so called; it is to be hoped that the efforts towards reform in the matter will not cease until completely successful.

Firing of Sawdust and Shavings.—"The air was forced into the furnace with the planer shavings at a velocity of about 12 feet per second, and at an average temperature of about 60 degrees Fahrenheit. The shavings were forced through a pipe 12 inches in diameter, above grate, into the combustion chamber. The pipe had a blast gate to regulate the air in order to maintain a pressure in the furnace, which a little more than balanced the ascending gases in the funnel or chimney. All the fireman had to do was to keep the furnace doors closed and watch the water in the gauges of his boiler. The combustion in the furnace was complete, as no smoke was visible. The shavings were forced into the combustion chamber in a spray-like manner, and were caught into a blaze the moment they entered. The oxygen of the air so forced into the furnace along with the shavings gave full support to the combustion.

FIRING WITH VARIOUS FUELS.

The amount of shavings consumed by being thus forced into the furnace was about fifty per cent. less than the amount consumed, when the fireman had to throw them in with his shovel."

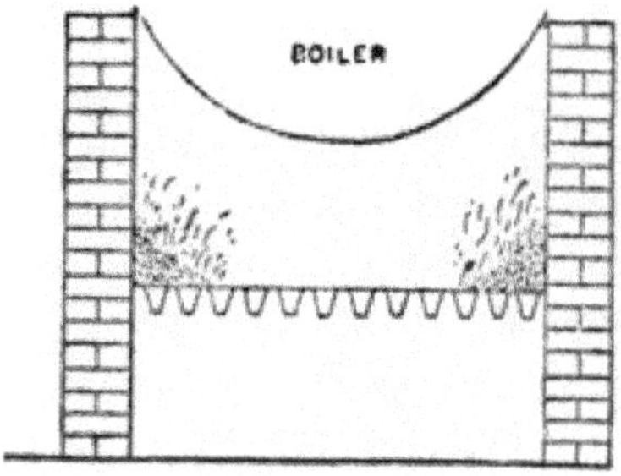

Fig. 9.

It is an important "point" when burning shavings or saw-dust with a blast, to keep the blower going without cessation, as there have been disastrous accidents caused by the flames going up the shutes, thence through the small dust tubes leading from the bin to the various machines.

In firing "shavings" by hand, it is necessary to burn them from the top as otherwise the fire and heat are only produced when all the shavings are charred. To do this, provide a a half inch gas pipe, to be used as a light poker; light the shaving fire, and when nearly burned take the half-inch pipe and divide the burning shavings through the middle, banking them against the side-walls as shown in Fig. 9. Now feed a pile of new shavings into the centre on the clean grate bars, as shown in Fig. 10, and close the furnace doors. The shavings will begin to burn from above, lighted from the two side fires, the air will pass through the bars into the shavings, where it will be heated and unite with the gas, making the combustion perfect, generating heat, and no smoke, and the fire will last much longer and require not half the labor in stoking.

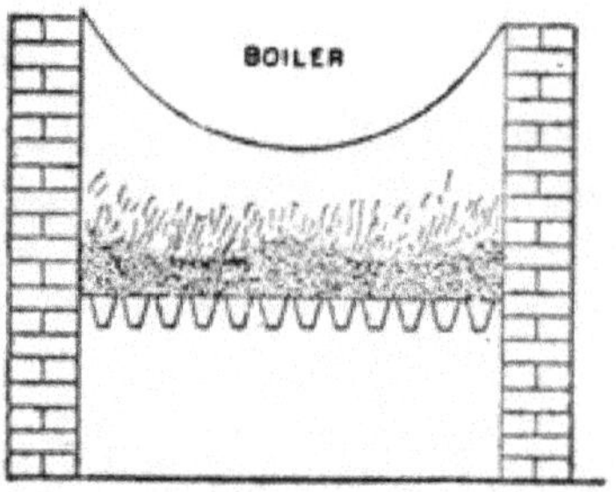

Fig. 10.

FIRING A LOCOMOTIVE.

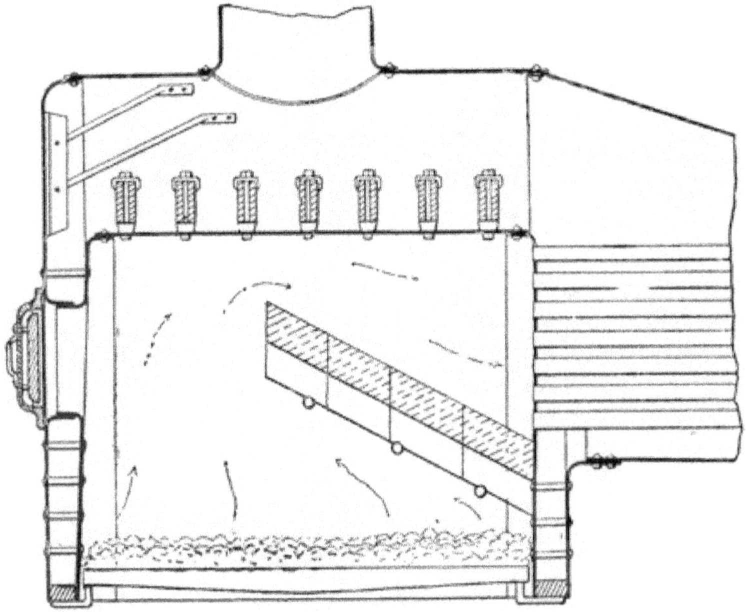

This figure exhibits the interior of the furnace of a locomotive engine, which varies greatly from the furnace of either a land or marine boiler. This difference is largely caused by the method of applying the draught for the air supply; in the locomotive this is effected by conducting the exhaust steam through pipes from the cylinders to the smoke-box and allowing it to escape up the smoke stack from apertures called exhaust nozzles; the velocity of the steam produces a vacuum, by which the products of combustion are drawn into the smoke-box with great power and forced out of the smoke stack into the open air.

To prevent the too quick passage of the gases into the flues an appliance called a fire brick arch has been adopted and has proved very efficient. In order to be self supporting it is built in the form of an arch, supported by the two sides of the fire box which serve for abutments. The arch has been sometimes replaced by a hollow riveted arrangement called a water table designed to increase the fire surface of the boiler.

FIRING A LOCOMOTIVE.

Firing a Locomotive.—No rules can possibly be given for firing a locomotive which would not be more misleading than helpful. This is owing to the great variations which exist in the circumstances of the use of the machine, as well as the differences which exist in the various types of the locomotive.

These variations may be alluded to, but not wholly described. 1. They consist of the sorts of fuel used in different sections of the country and frequently on different ends of the same railroad; hard coal, soft coal, and wood all require different management in the furnace. 2. The speed and weight of the train, the varying number of cars and frequency of stopping places, all influence the duties of the fireman and tax his skill. 3. The temperature of the air, whether cold or warm, dry weather or rain, and night time and day time each taxes the skill of the fireman.

Hence, to be an experienced fireman in one section of the country and under certain circumstances does not warrant the assurance of success under other conditions and in another location. The subject requires constant study and operation among not only "new men" but those longest in the service.

More than in any other case to be recalled, must the fireman of a locomotive depend upon the personal instruction of the engineer in charge of the locomotive.

Firing with Tan Bark.—Tan bark can be burned upon common grates and in the ordinary furnace by a mixture of bituminous screenings. One shovel full of screenings to four or five of bark will produce a more economical result than the tan bark separate, as the coal gives body to the fire and forms a hot clinker bed upon which the bark may rest without falling through the spaces in the grate bars, and with the coal, more air can be introduced to the furnace.

The above relates to common furnaces, but special fire boxes have been recently put into operation, fed by power appliances, which work admirably. The "point" principally to be noted as to the efficacy of tan bark as a fuel, is to the effect, that like peat, the drier it is the more valuable is it as a fuel.

POINTS RELATING TO FIRING.

THE PROCESS OF BOILING. Let it be remembered that the boiling spoken of so often is really caused by the formation of the steam particles, and that without the boiling there can be but a very slight quantity of steam produced.

While pure water boils at 212°, if it is saturated with common salt, it boils only on attaining 224°, alum boils at 220°, sal ammoniac at 236°, acetate of soda at 256°, pure nitric acid boils at 248°, and pure sulphuric acid at 620°.

ON THE FIRST APPLICATION OF HEAT to water small bubbles soon begin to form and rise to the surface; these consist of air, which all water contains dissolved in it. When it reaches the boiling point the bubbles that rise in it are principally steam.

IN THE CASE OF A NEW PLANT, or where the boiler has some time been idle it is frequently advisable to build a smal' fire in the base of the chimney before starting the boiler fires This will serve to heat the chimney and drive out any moist ure that may have collected in the interior and will frequently prevent the disagreeable smoking that often follows the building of a fire in the furnace.

ALWAYS BEAR IN MIND that the steam in the boilers and engines is pressing outward on the walls that confine it in every direction; and that the enormous forces you are handling, warn you to be careful.

When starting fires close the gauge cocks and safety valve as soon as steam begins to form.

GO SLOW. It is necessary to start all new boilers very slowly. The change from hot to cold is an immense one in its effects on the contraction and expansion of the boiler, the change of dimension by expansion is a force of the greatest magnitude and cannot be over-estimated. Leaks which start in boilers that were well made and perfectly tight can be attributed to this cause. Something must give if fires are driven on the start, and this entails trouble and expense that there is no occasion for. This custom applies to engines and steam pipes as well as to boilers. No one of any experience will open a stop valve and let a full head of live steam into a cold line of pipe or a cold engine.

POINTS RELATING TO FIRING.

To preserve the grate bars from excessive heat, when first firing a boiler, it is well to sprinkle a thin layer of coal upon the grates before putting in the shavings and wood for starting the fire. This practice tends greatly to prolong the life of the grate-bars.

The fuel should generally be dry when used. Hard coal, however, may be dampened a little to good advantage, as it is then less liable to crowd and will burn more freely.

Air, high temperature, and sufficient time are the principal points in firing a steam boiler.

In first firing up make sure that the throttle valve is closed, in order that the steam first formed may not pass over into the engine cylinder and fill it with water of condensation. If the throttle valve leak steam it should be repaired at the first opportunity.

Keep all heating surfaces free from soot and ashes.

Radiant rays go in all directions, yet they act in the most efficient manner when striking a surface exactly at a right angle to their line of movement. The sides of a fire-box are for that reason not as efficient as the surface over the fire, and a flat surface over the fire is the best that can be had, so far as that fact alone is considered.

When combustion is completed in a furnace then the balance of the boiler beyond the bridge wall can be utilized for taking up heat from the gases. The most of this heat has to be absorbed by actual contact; thus by the tubes the gases are finely divided, allowing that necessary contact.

Combustion should be completed on the grates for the reason that it can be effected there at the highest temperature. When this is accomplished, the fullest benefit is had from radiant heat striking the bottom of the boiler—*it is just there that the bulk of the work is done.*

There must necessarily be some waste of heat by its passing up the chimney to maintain draft. It is well to have the gases, as they enter the chimney, as much below 600 deg. F. (down to near the temperature of the steam) as you can and yet maintain perfect combustion.

Every steam engine has certain well-defined sounds in action which we call noises, for want of a better term, and it is upon them and their continuance that an engineer depends for assurance that all is going well.

This remark also applies to the steam boiler, which has, so to speak, a language of its own, varying in volume from the slight whisper which announces a leaking joint to the thunder burst which terribly follows a destructive explosion. The hoarse note of the safety valve is none the less significant because common.

The dampers and doors to the furnace and ash pit should always be closed after the fire has been drawn, in order to keep the heat of the boiler as long as possible.

But the damper must never be entirely closed while there is fire on the grate as explosions dangerous in their character might occur in the furnace from the accumulated gases.

Flues or tubes should often be swept, as soot, in addition to its liability to becoming charged with a corroding acid, is a non-conductor of heat, and the short time spent in cleaning them will be repaid by the saving of labor in keeping up steam. In an establishment where they used but half a ton of bituminous coal per day, the time of raising steam in the morning was fifty per cent. longer when the tubes were unswept for one week than when they were swept three times a week.

SMOKE will not be seen *if combustion is perfect.* Good firing will abate most of the smoke.

Coals, at the highest furnace temperature, radiate much heat, whereas gases ignited at and beyond the bridge wall radiate comparatively little heat—it is a law in nature for a solid body highly heated to radiate heat to another solid body.

DRY AND CLEAN is the condition in which the boiler should be kept, i. e., dry outside and clean both inside and out.

POINTS RELATING TO FIRING

To haul his furnace fire and open the safety valve before seeking his own safety or the preservation of property, is the duty of the fireman in the event of fire threatening to burn a whole establishment.

Many, now prominent, engineers have made their first reputation by remembering to do this at a critical time.

WHEN WATER IS PUMPED into the boiler or allowed to run in, some opening must be given for the escape of the contained air, usually the most convenient way is to open the upper guage cock after the fire has been lighted until cloudy steam begins to escape.

In a summary of experiments made in England, it is stated that:—

"A moderately thick and hot fire with rapid draft uniformly gave the best results."

"Combustion of black smoke by additional air was a loss."

"In all experiments the highest result was always obtained when all the air was introduced through the fire bars."

"Difference in mode of firing only may produce a difference of 13 per cent." (in economy).

The thickness of the fire under the boiler should be in accordance with the quality and size of the fuel. For hard coal the fire should be as thin as possible, from three to six inches deep; when soft coal is used, the fire should be thicker, from five to eight inches deep.

If it is required to burn coal dust without any change of grates, wetting the coal is of advantage; not that it increases its heat power, but because it keeps it from falling through the grates or going up the chimney. The same is true of burning shavings; by watering they are held in the furnace, and the firing is done more easily and with better results.

STIRRING THE FIRE should be avoided as much as possible; firing should be performed evenly and regularly, a little at a time, as it causes waste fuel to disturb the combustion and by making the fuel fall through the grates into the ash pit; hence do not "clean" fires oftener then absolutely necessary.

POINTS RELATING TO FIRING.

The slower the velocity of the gases before they pass the damper, the more nearly can they be brought down to the temperature of the steam, hence with a high chimney and strong draft the dampers should be kept nearly closed, if the boiler capacity will permit it.

No arbitrary rule can be laid down for keeping fires thick or thin. Under some conditions a thin fire is the best, under others a thick fire gives best economy. This rule, however, governs either case : you must have so active a fire as to give strong radiant heat.

One of the highest aims of an expert fireman should be to keep the largest possible portion of his grate area in a condition to give great radiant heat the largest possible part of a day—using anthracite coal by firing light, quick and often, not covering all of the incandescent coals. Using bituminous coal, hand firing, by coking it *very near* the dead plate, allowing some air to go through openings in the door, and by pushing toward the bridge wall only live coals—when slicing, to open the door only far enough to work the bar ; this is done with great skill in some cases.

REGULATING THE DRAFT.—This should be done so as to admit *the exact quantity of air* into the furnace, neither too much nor too little. It should be remembered that fuel cannot be burned without air and if too much air is admitted it cools the furnace and checks combustion. It is a good plan to decrease the draft when firing or cleaning out, by partly closing the damper or shutting off the air usually admitted from below the grates ; this is to have just draught enough to prevent the flame from rushing out when the door is opened.

By luminous flame is generally meant that which burns with a bright yellow to white color. All flame under a boiler is not luminous, sometimes the whole or a part of it will be red or blue. The more luminous the flame, that is to say, the nearer white it is, the better combustion.

RULES RELATING TO FIRING.

To DETERMINE THE TEMPERATURE OF A FURNACE FIRE the following table is of use. The colors are to be observed and the corresponding degrees of heat will be approximately as follows:

Faint red	960° F.
Bright red	1,300° F.
Cherry red	1,600° F.
Dull orange	2,000° F.
Bright orange	2,100° F.
White heat	2,400° F.
Brilliant white heat	2,700° F.

That is to say, when the furnace is at a "white heat" the heat equals 2,400 degrees Fahrenheit, etc.

Another method of finding the furnace heat is by submitting a small portion of a particular metal to the heat.

Tin melts at	442° F.	
Lead " "	617° F.	
Zinc " "	700° F.	nearly.
Antimony melts at	810 to 1,150° F.	
Silver melts at	1,832 to 1,873° F.	
Cast Iron melts at	2,000° F.	nearly.
Steel " "	2,500° F.	"
Wrought Iron melts at	2,700° F.	"
Hammered Iron melts at	2,900° F.	"

FOAMING IN BOILERS.

The causes are—dirty water, trying to evaporate more water than the size and construction of the boiler is intended for, taking the steam too low down, insufficient steam room, imperfect construction of boiler, too small a steam pipe and sometimes it is produced by carrying the water line too high.

Too little attention is paid to boilers with regard to their evaporating power. Where the boiler is large enough for the water to circulate, and there is surface enough to give off the steam, foaming never occurs.

As the particles of the steam have to escape to the surface of the water in the boiler, unless that is in proportion to the amount of steam to be generated, it will be delivered with such violence that the water will be mixed with it, and cause foaming.

FOAMING IN BOILERS.

For violent ebollition a plate hung over the hole where the steam enters the dome from the boiler, is a good thing, and prevents a rush of water by breaking it, when the throttle is opened suddenly.

In cases of very violent foaming it is imperative to check the draft and cover the fires.

The steam pipe may be carried through the flange six inches into the dome—which will prevent the water from entering the pipes by following the sides of the dome as it does.

A similar case of priming of the boilers of the U. S. Steamer Galena was stopped by removing some of the tubes under the smoke stack, and substituting bolts.

Clean water, plenty of surface, plenty of steam room, large steam pipes, boilers large enough to generate steam without forcing the fires, are all that is required to prevent foaming.

A high pressure insures tranquillity at the surface, and the steam itself being more dense it comes away in a more compact form, and the ebullition at the surface is no greater than at a lower pressure. When a boiler foams it is best usually to close the throttle to check the flow, and that keeps up the pressure and lessens the sudden delivery.

Too many flues in a boiler obstruct the passage of the steam from the lower part of the boiler on its way to the surface—this is a fault in construction.

An engineer who had been troubled with priming, finally removed 36 of the tubes in the centre of the boiler, so as to centralize the heating effect of the fire, thereby increasing the rapidity of ebullition at the centre, while reducing it at the circumference. The effect of the change was very marked. The priming disappeared at once. The water line became nearly constant, the extreme variation being reduced to two inches.

A CHAPTER OF DON'TS.

Which is another way of repeating what has already been said.

1. ***Don't*** empty the boiler when the brick work is hot.
2. ***Don't*** pump cold water into a hot boiler.
3. ***Don't*** allow filth of any kind to accumulate around the boiler or boiler room.
4. ***Don't*** leave your shovel or any other tool out of its appointed place when not in use.
5. ***Don't*** fail to keep all the bright work about the boiler neat and "shiny."
6. ***Don't*** forget that negligence causes great loss and danger.
7. ***Don't*** fail to be alert and ready-minded and ready-headed about the boiler and furnace.
8. ***Don't*** read newspapers when on duty.
9. ***Don't*** fire up too quickly.
10. ***Don't*** let any water or dampness come on the outside of your boiler.
11. ***Don't*** let any dampness get into the boiler and pipe coverings.
12. ***Don't*** fail to see that you have plenty of water in the boiler in the morning.
13. ***Don't*** fail to keep the water at the same height in the boiler all day.
14. ***Don't*** let any one talk to you when firing.
15. ***Don't*** allow water to remain on the floor about the boiler.
16. ***Don't*** fail to blow off steam once or twice per day according as the water is more or less pure.
17. ***Don't*** fail to close the blow-off cock, when blowing off, when the water in the boiler has sunk to one and a half inches.
18. ***Don't*** fail, while cleaning the boiler, to examine and clean all cocks, valves and pipes and look to all joints and packings.

A CHAPTER OF DON'TS.

19. ***Don't*** commence cleaning the boiler until it has had time to cool.

20. ***Don't*** forget daily to see that the safety valve moves freely and is tight.

21. ***Don't*** fail to clean the boiler inside frequently and carefully.

22. ***Don't*** fail to notice that the steam gauge is in order.

23. ***Don't*** fail to keep an eye out for leaks and have them repaired immediately, no matter how small.

24. ***Don't*** fail to empty the boiler every week or two and re-fill it with fresh water.

25. ***Don't*** let any air into the furnace, except what goes through the grate bars, or the smoke burners, so called, by which the air is highly heated.

26. ***Don't*** increase the load on the safety valve beyond the pressure allowed by the inspector.

27. ***Don't*** fail to open the doors of the furnace and start the pump when the pressure is increased beyond the amount allowed, *but*

28. ***Don't*** fail to draw the fires *when there is danger* from the water having fallen too low.

29. ***Don't*** fail to check the fire—if too hot to draw, do it with fresh coal, damp ashes, clinkers or soil; *and*

30. ***Don't*** fail to open the doors of the furnace and close the ash pit doors at the time the fire is checked—*and*

31. ***Don't*** decrease the steam pressure by feeding in water or suddenly blowing off steam, *and*

32. ***Don't*** touch the safety valve, even if it be opened or closed, *and*

33. ***Don't*** change the feed apparatus if it is working, or the throttle-valve be open; let them both remain as they are for a short time, *and*

34. ***Don't*** fail to change them very cautiously and slowly when you close them, and

35. ***Don't*** fail to be very cool and brave while resolute in observing these last seven "Dont's."

A CHAPTER OF DONT'S.

36. ***Don't*** fail to keep yourself neat and tidy.

37. ***Don't*** fail to be polite as well as neat and brave.

38. ***Don't*** fail to keep the tubes clear and free from soot and ashes.

39. ***Don't*** let too many ashes gather in the ashpit.

40. ***Don't*** disturb the fire when it is burning good nor stir it up too often.

41. ***Don't*** be afraid to get instruction from books and engineering papers.

42. ***Don't*** fail to make an honest self-examination as to points upon which you may be ignorant, and really need to know in order to properly attend to your duties.

43. ***Don't*** allow too much smoke to issue from the top of the chimney if the cause lies within your power to prevent it.

44. ***Don't*** think that after working at firing and its kindred duties for a year or two that *the whole subject* of engineering has been learned.

45. ***Don't*** forget that one of the best helps in getting forward is the possession of a vigorous and well balanced mind and body—this covers temperance and kindred virtues and a willingness to acquire and impart knowledge.

46. ***Don't*** forget to have your steam gauge tested at least once in three months.

47. ***Don't*** use a wire or metallic rod as a handle to a swab in cleaning the glass tube of a water gauge for the glass may suddenly fly to pieces when in use within a short time afterwards.

48. ***Don't*** forget that steam pumps require as much attention as a steam engine.

49. ***Don't*** run a steam pump piston, unless in an emergency, at a speed exceeding 80 to 100 feet per minute.

50. ***Don't*** do anything without a good reason for it about the engine or boiler, but when you are obliged to do anything, do it thoroughly and as quickly as possible.

A CHAPTER OF DONT'S.

51. ***Don't*** forget to sprinkle a thin layer of coal on the grates before lighting the shavings and wood in the morning. This practice preserves the grate bars.

52. ***Don't*** don't take the cap off a bearing and remove the upper brass simply to see if things are working well ; if there is any trouble it will soon give you notice, and, besides, you never can replace the brass in exactly its former position, so that you may find that the bearing will heat soon afterwards, owing to your own uncalled-for interference.

53. ***Don't*** put sulphur on a hot bearing, unless you intend to ruin the brasses.

54. ***Don't*** use washed waste that has a harsh feel, as the chemicals used in cleansing it have not been thoroughly removed.

55. ***Don't,*** in case of an extensive fire, involving the whole business, rush off without drawing the fires, and raising and *propping open* the safety valve of the boiler.

56. ***Don't*** fail to preserve your health, for "a sound mind in a sound body" is beyond a money valuation.

57. ***Don't*** fail to remember that engineers and firemen are in control of the great underlying force of modern civilization; hence, to do nothing to lower the dignity of the profession.

58. ***Don't*** forget that in the care and management of the steam boiler the first thing required is an unceasing watchfulness —*watch-care.*

59. ***Don't*** forget that an intemperate, reckless or indifferent man has no business in the place of trust of a steam boiler attendant.

60. ***Don't*** allow even a day to pass without adding one or more facts to your knowledge of engineering in some of its branches.

STEAM GENERATORS.

In the examinations held by duly appointed officers to determine the fitness of candidates for receiving an engineer's license the principal stress is laid upon the applicant's knowledge of the parts and true proportions of the various designs of steam boilers, and his experience in managing them.

In fact, if there were no boilers there would be no examinations, as the laws are framed, certificates issued and steam boiler inspection companies formed to assure the public safety in life, limb and property, from the dangers arising from so-called mysterious boiler explosions.

Hence an almost undue proportion of engineer's examinations are devoted to the steam boiler, its management and construction. But the subject is worthy of the best and most thoughtful attention. Every year adds to the number of steam boilers in use. With the expanding area and growth of population, the number of steam plants are multiplied and in turn each new steam boiler demands a careful attendant.

There is this difference between the boiler and the engine. When the latter is delivered from the shop and set up, it does its work with an almost unvarying uniformity, while the boiler is a constant care. It is admitted that the engine has reached a much greater state of perfection and does its duty with very much more reliability than the boiler.

Even when vigilant precautions are observed, from the moment a steam boiler is constructed until it is finally destroyed there are numerous insidious agents perpetually at work which tend to weaken it. There is nothing from which the iron can draw sustenance to replace its losses. The atmosphere without and the air within the boiler, the water as it

STEAM GENERATORS.

cators through the feed-pipe and containing mineral and organic substances, steam into which the water is converted, the sediment which is precipitated by boiling the water, the fire and the sulphurous and other acids of the fuel, are all natural enemies of the iron; they sap its strength, not only while the boiler is at work and undergoing constant strain, but in the morning before fire is started, and at noon, night, Sundays, and other holidays it is preyed upon by these and other corroding agents.

These are the reasons which impress the true engineer with a constant solicitude regarding the daily and even momentary action of the steam generator.

Description.

The Steam Boiler in its simplest form was simply a closed vessel partly filled with water and which was heated by a fire box, but as steam plants are divided into two principal parts, the engine and the boiler, so the latter is divided again into the furnace and boiler, each of which is essential to the other. The furnace contains the fuel to be burnt, the boiler contains the water to be evaporated.

There must be a steam space to hold the steam when generated; heating surface to transmit the heat from the burning fuel to the water; a chimney or other apparatus to cause a draught to the furnace and to carry away the products of combustion; and various fittings for supplying the boiler with water, for carrying away the steam when formed to the engine in which it is used; for allowing steam to escape into the open air when it forms faster than it can be used; for ascertaining the quantity of water in the boiler, for ascertaining the pressure of the steam, etc., all of which, together with the engine and its appliances is called A STEAM PLANT.

The forms in which steam generators are built are numerous, but may be divided into three classes, viz.: stationary, locomotive and marine boilers, which terms designate the uses for which they are intended; in this work we have to deal mainly with the first-named, although a description with illustration is given of each type or form.

AN UPRIGHT STEAM BOILER.

To illustrate the operations of a steam generator, we give the details of an appliance, which may be compared to the letter A of the alphabet, or the figure 1 of the numerals, so simple is it.

Fig. 11, is an elevation of boiler, fig. 12 a vertical section through its axis, and fig. 13 a horizontal section through the furnace bars.

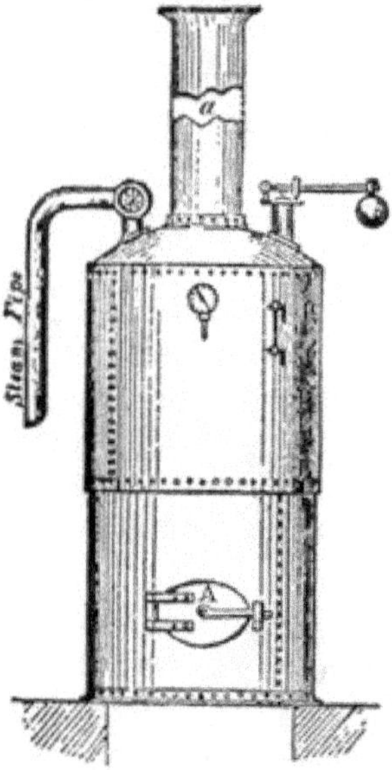

Fig. 11.

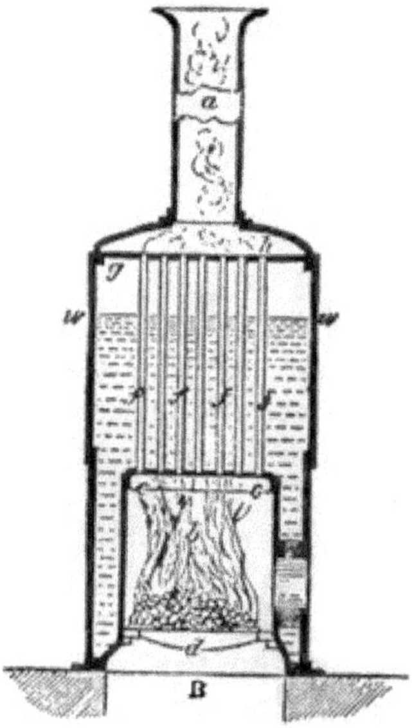

Fig. 12.

The type of steam generator here exhibited is what is known as a vertical tubular boiler. The outside casing or shell is cylindrical in shape, and is composed of iron or steel plates riveted together. The top, which is likewise composed of the same plates is slightly dome-shaped, except at the center, which is away in order to receive the chimney *a*, which is round in shape and formed of thin wrought iron plates. The interior is shown in vertical section in fig. 12. It consists of a furnace chamber, *b*, which contains the fire. The furnace is formed like the shell of the boiler of wrought iron or steel plates by flanging and riveting. The bottom is occupied by the grating, on which rests the incandescent fuel. The grating consists of

UPRIGHT STEAM BOILERS.

a number of cast-iron bars, *d* (fig. 12), and shown in plan in fig. 13, placed so as to have interstices between them like the grate of an ordinary fireplace. The bottom of the furnace is firmly secured to the outside shell of the boiler in the manner shown in fig. 12. The top covering plate *cc*, is perforated with a number of circular holes of from one and a half to three inches diameter, according to the size of the boiler. Into each of these holes is fixed a vertical tube made of brass, wrought iron, or steel, shown at *fff* (fig. 12). These tubes pass through similar holes, at their top ends in the plate *g*, which latter is firmly riveted to the outside shell of the boiler. The tubes are also firmly attached to the two plates, *cc, g*. They serve to convey the flame, smoke, and hot air from the fire to the smoke box, *h*, and the chimney, *a*, and at the same time their sides provide ample heating surface to allow the heat contained in the products of combustion to escape into the water. The fresh fuel is thrown on the grating when required through the fire door, A (fig. 11). The ashes, cinders, etc., fall between the fire bars into the ash pit, B (fig. 12). The water is contained in the space between the shell of the boiler, the furnace chamber, and the tubes. It is kept at or about the level, *ww* (fig. 12), the space above this part being reserved for the steam as it rises. The heat, of course, escapes into the water, through the sides and top plate of the furnace, and through the sides of the tubes. The steam which, as it rises from the boiling water, ascends into the space above *ww*, is thence led away by the steam pipe to the engine. Unless consumed quickly enough by the engine, the steam would accumulate too much within the boiler, and its pressure would rise to a dangerous point. To provide against this contingency the steam is enabled to escape when it rises above a certain pressure through the safety valve, which is shown in sketch on the top of the boiler in fig. 11. The details of the construction of safety valves will be found fully described in another section of this work, which is devoted exclusively to the consideration of boiler fittings. In the same chapters will be found full descriptions of the various fittings and accessories of boilers, such as the water and pressure gauges, the apparatus for feeding the boiler with water, for producing the requisite draught of air to maintain the combustion, and also the particulars of the construction of the boilers themselves and their furnaces.

Fig. 13.

THE GROWTH OF THE STEAM BOILER.

After the first crude forms, such as that used in connection with the Baranca and Newcoman engine, and numerous others. The steam boiler which came into very general use was *the plain cylinder boiler.* An illustration is given of this in figures 14 and 15.

It consists of a cylinder A, formed of iron plate with hemispherical ends B.B. set horizontally in brick work C. The lower part of this cylinder contains the water, the upper part the steam. The furnace D is outside the cylinder, being beneath one end; it consists simply of grate bars *e e* set in the brick work at a convenient distance below the bottom of the boiler.

The sides and front of the furnace are walls of brick work, which, being continued upwards support the end of the cylinder. The fuel is thrown on the bars through the door which is set in the front brick work. The air enters between

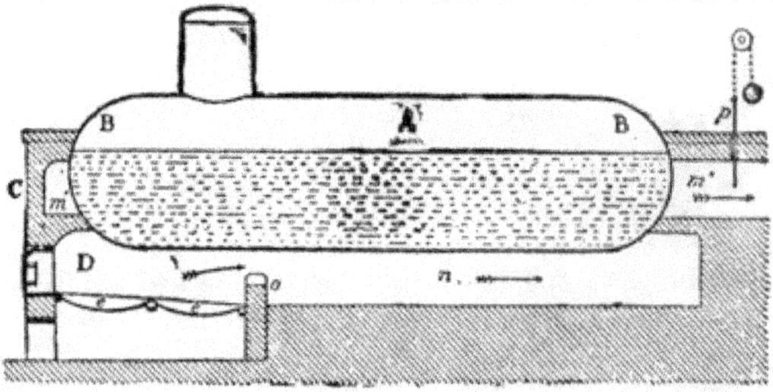

Fig. 14.

the grate bars from below. The portion below the bars is called the ash pit. The flame and hot gasses, when formed, first strike on the bottom of the boiler, and are then carried forward by the draft, to the so-called bridge wall *o*, which is a projecting piece of brick work which contracts the area of the flue *n* and forces all

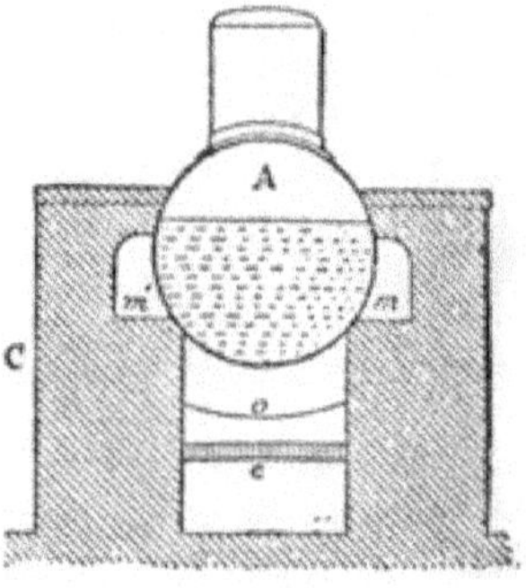

Fig. 15.

THE GROWTH OF THE STEAM BOILER.

the products of combustion to keep close to the bottom of the boiler.

Thence the gasses pass along the flue *n*, and return part one side of the cylinder in the flue *m* (fig. 15) and back again by the other side flue *m* to the far end of the boiler, whence they escape up the chimney. This latter is provided with a door or damper *p*, which can be closed or opened at will, so as to regulate the draught.

This boiler has been in use for nearly one hundred years, but has two great defects. The first is that the area of heating surface, that is the parts of the boiler in contact with the flames, is too small in proportion to the bulk of the boiler; the second is, that if the water contains solid matter in solution, as nearly all the water does to a greater or less extent, this matter becomes deposited on the bottom of the boiler just where the greatest evaporation takes place. The deposit, being a non-conductor, prevents the heat of the fuel from reaching the water in sufficient quantities, thus rendering the heating surface inefficient; and further, by preventing the heat from escaping to the water, it causes the plates to become unduly heated, and quickly burnt out.

There is another defect belonging to this system of boiler to which many engineers attach great importance, viz.: that the temperature in each of the three flues *n*, *m*, *m'* is very different, and consequently that the metal of which the shell of the boiler is composed expands very unequally in each of the flues, and cracks are very likely to take place when the effects of the changes of temperature are most felt. It will be noted that the flames and gasses in this earliest type of steam boiler make three turns before reaching the chimney, and as these boilers were made frequently as much as 40 feet long it gave the extreme length of 120 feet to the heat products.

THE GROWTH OF THE STEAM BOILER.

THE CORNISH BOILER is the next form in time and excellence. This is illustrated in figures 16 and 17.

It consists also of a cylindrical shell *A*, with flat ends as exhibited in cuts. The furnace, however, instead of being situated underneath the front end of the shell, is enclosed in it in a second cylinder *B*, having usually a diameter a little greater than half that of the boiler shell. The arrangement of the grate and bridge is evident from the diagram. After passing the bridge wall the heat products travel along through the internal cylinder *B*, till they reach the back end of the boiler; then return to the front again, by the two side flues *m, m,'* and thence back again to the chimney by the bottom of flue *n*.

In this form of boiler the heating surface exceeds that of the last described by an amount equal to the area of the internal flues, while the internal capacity is diminished by its cubic contents; hence for boilers of equal external dimensions, the ratio of heating surface to mass of water to be heated, is greatly increased. Boilers of this sort can, however, never be made of

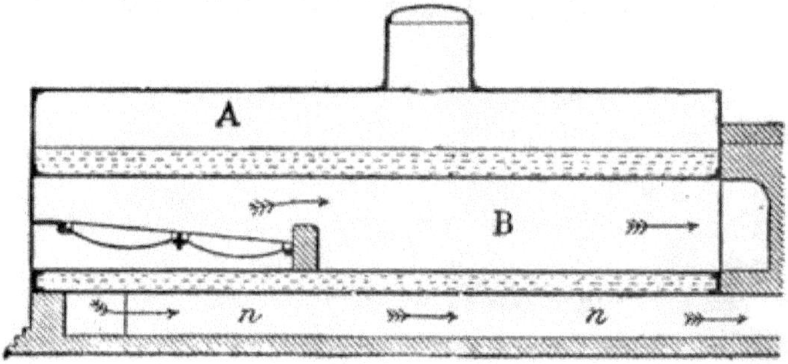

Fig. 16.

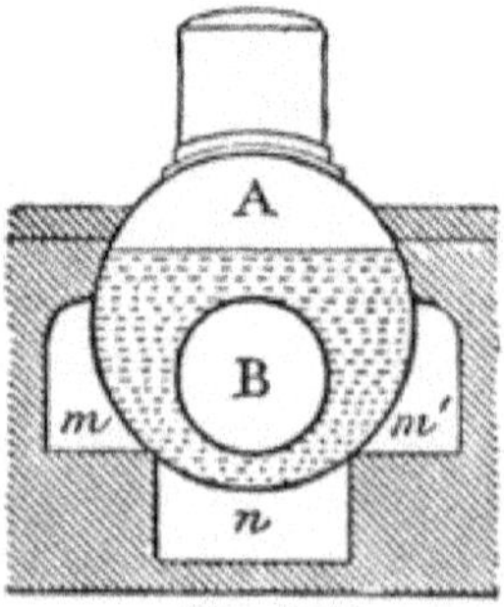

Fig. 17.

as small diameters as the plain cylindrical sort, on account of the necessity of finding room inside, below the water level, for the furnace and flue.

The disadvantage, too, of the deposits mentioned in the plain cylinder is, to a great extent got over in the Cornish boiler, for the

THE GROWTH OF THE STEAM BOILER.

bottom, where the deposit chiefly takes place, is the coolest instead of being the hottest part of the heating surface.

But the disadvantage of unequal expansion also exists in this type of boiler, as the internal flue in the Cornish system is the hottest portion of the boiler, and consequently undergoes a greater lengthways expansion than the flues. The result is to bulge out the ends, and when the boiler is out of use, the flue returns to its regular size, and thus has a tendency to work loose from the ends to which it is riveted and if the ends are too rigid to move, a very serious strain comes on the points of the flue.

Even while in use the flue of a Cornish boiler is liable to undergo great changes in temperature, according to the state of the fire; when this latter is very low, or when fresh fuel has been thrown on, the temperature is a minimum and reaches a maximum again when the fresh fuel commences to burn fiercely. This constant expansion and contraction is found in practice to also so weaken the tube that it frequently collapses or is pressed together, resulting in great disaster.

This led to the production and adoption of the—

LANCASHIRE BOILER, contrived to remedy this inconvenience and also to attain a more perfect combustion, the arrangement of the furnaces of which is shown in fig. 19 and fig. 20.

It will be observed that there are two internal furnaces instead of one, as in the Cornish type. These furnaces are sometimes each continued as a separate flue to the other end of the boiler as shown in the cuts; but as a rule they emerge into one internal flue. They are supposed to be fired alternately, and the smoke and unburned gases issuing from the fresh fuel are ignited in the flue by the hot air proceeding from the other furnace, the fuel in which is in a state of incandescence. Thus all violent changes in the temperature are avoided, and the waste of fuel due to unburned gases is avoided, if the firing is properly conducted.

The disadvantage of the Lancashire boiler is the difficulty of finding adequate room for the two furnaces without unduly in-

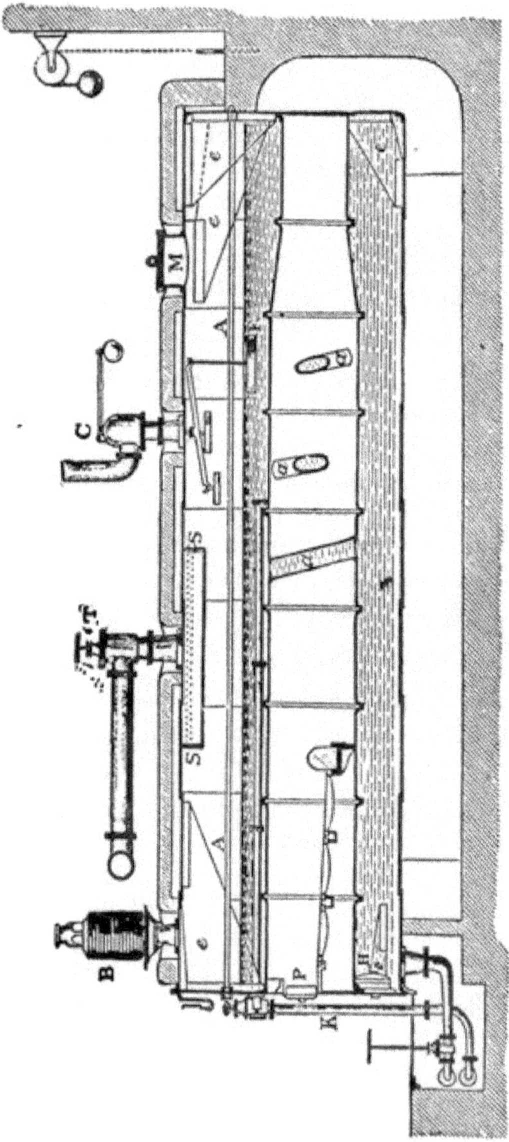

LANCASHIRE BOILER.—Fig. 18.

creasing the diameter of the shell. Low furnaces are extremely unfavorable to complete combustion, the comparatively cold crown plates, when they are in contact with the water of the boiler, extinguishing the flames from the fuel, when they are just formed, while the narrow space between the fuel and the crown does not admit the proper quantity of air being supplied above the fuel to complete the combustion of the gases, as they arise.

On the other hand, though this boiler favors the formation of the smoke, it supplies the means of completing the combustion afterwards, as before explained, by means of the hot air from the second furnace.

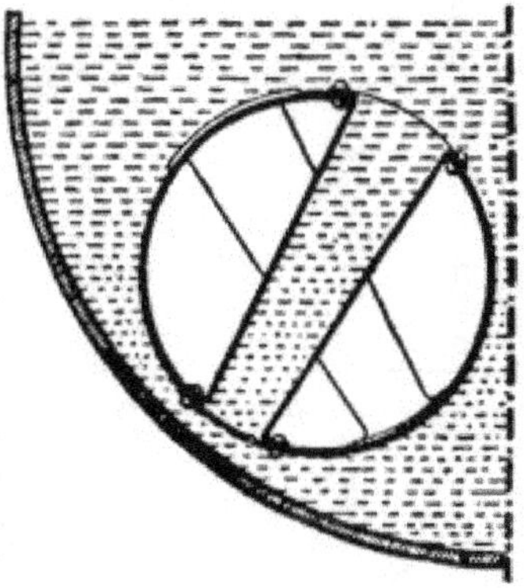

Another disadvantage is the danger of collapsing the internal flue already spoken of; this is remedied by the introduction of what are called the galloway tubes, illustrated in the cut shown on this page, which is a cross section of the water tubes shown in Figs. 18 and 20.

These tubes not only contribute to strengthen the flues but they add to the heating surface and greatly promote the circulation so important in the water space.

NOTE.

These descriptions and illustrations of the Lancashire boiler are of general value, owing to the fact that very many exhaustive tests and experiments upon steam economy have been made and permanently recorded in connection with this form of steam generator.

THE GROWTH OF THE STEAM BOILER.

In the GALLOWAY form of boiler the flue is sustained and stiffened by the introduction of numerous conical tubes, flanged at the two ends and riveted across the flue. These tubes, a sketch of which are given in fig. 18 (*a*), are in free communication with the water of the boiler, and besides acting as stiffeners, they also serve to increase the heating surface and to promote circulation.

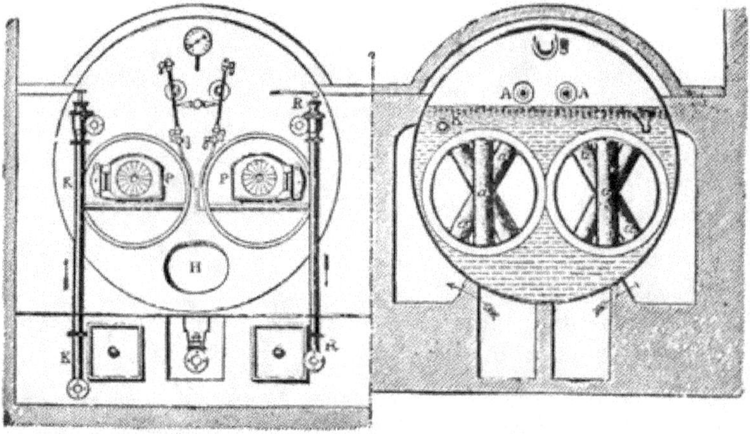

Fig. 19. Fig. 20.

The illustration (figs. 18, 19 and 20) give all the principal details of a Lancashire boiler fitted with Galloway tubes. Fig. 18 represents a longitudinal section and figs. 19 and 20 shows on a large scale an end view of the front of the boiler with its fittings and also a transverse section. The arrangement of the furnaces, flues, and the Galloway tubes is sufficiently obvious from the drawings. The usual length of these boilers is 27 feet, though they are occasionally made as short as 21 feet.

The minimum diameter of the furnaces is 33 inches, and in order to contain these comfortably the diameter of the boiler should not be less than 7 feet. The ends of the boiler are flat, and are prevented from bulging outwards by being held in place by the furnaces and flues which stay the two ends together and also by the so-called gusset stays *e, e*. In addition to the latter the flat ends of the boiler have longitudinal rods to tie them together; one of these is shown at *A, A*. fig. 18.

THE GROWTH OF THE STEAM BOILER.

The steam is collected in the pipe *S*, which is perforated all along the top so as to admit the steam and exclude the water spray which may rise to the surface during ebullition. The steam thence passes to the stop valve *T* outside the boiler and thence to the steam pipes to the engines.

There are two safety valves on top of the boiler on *B* (fig. 18), being of the dead weight type described hereafter, and the other, *C*, being a so-called low water safety valve. It is attached by means of a lever and rod to the float *F*, which ordinarily rests on the surface of the water. When through any neglect, the water sinks below its proper level the float sinks also, causing the valve to open, thus allowing steam to escape and giving an alarm. *M* is the manhole with its covering plate, which admits of access to the interior of the boiler and *H* is the mud hole by which the sediment which accumulates all along the bottom is raked out. Below the front end and underneath, the pipe and stay valve are shown, by which the boiler can be emptied or blown off.

On the front of the boiler (fig 19) are shown, the pressure gauges, the water gauges and the furnace door ; *K* is the feed pipe ; *R*, *R*, a pipe and cock for blowing off steam. In the front of the setting are shown two iron doors by which access may be gained to the two lower external flues for cleaning purposes.

In the Lancashire boiler it is considered advisable to take the products of combustion, after they leave the internal flues, along the bottom of the boiler, and then back to the chimney by the side. When this plan is adopted the bottom is kept hotter than would otherwise be the case, and circulation is promoted, which prevents the coldest water from accumulating at the bottom.

The Galloway (or Lancashire) boiler is considered the most economical boiler used in England, and is being introduced into the United States with success. The long traverse of heat provided (three turns of about 27 feet each) contributes greatly to its efficiency.

THE GROWTH OF THE STEAM BOILER.

It may be useful to add the following data relating to this approved steam generator, being the principal dimensions and other data of the boiler shown in fig. 18:

Steam pressure, 75 lbs. per sq. inch.		
Length, 27 feet.	Heating surface:	
Diameter, 7 feet.	In furnace and flues	450 sq. feet.
Weight, total, $15\frac{1}{2}$ tons.	In Galloway pipes,	30 "
Shell plates, $\frac{7}{16}$ inch.	In external flues,	370 "
Furnace diameter, 33 inches.		——
Furnace Plates, $\frac{3}{8}$ inch.		850 sq. feet.
End plates, $\frac{1}{2}$ inch.		
Grate area, 33 sq. feet.		

We have thus detailed step by step the improvement of the steam boiler to a point where it is necessary to describe at length the locomotive, the marine, the horizontal tubular and the water tube boilers, which four forms comprehend ninety-nine out of one hundred steam generators in use in the civilized world at the present time.

MARINE BOILERS.

The boilers used on board steamships are of two principal types. The older sort used for steam of comparatively low temperature, viz.: up to 35 lbs. per square inch, is usually made of flat plates stayed together, after the manner of the exterior and interior fire boxes of a locomotive boiler.

Medium high pressure marine boilers, constructed for steam of 60 to 150 lbs. per square inch, are circular or oval in cross section, and are fitted with round interior furnaces and flues like land boilers. There are many variations of marine boilers, adapted to suit special circumstances. Fig. 22 shows a front elevation and partial sections of a pair of such boilers and fig. 23 shows one of them in longitudinal vertical section.

THE MARINE STEAM BOILER.

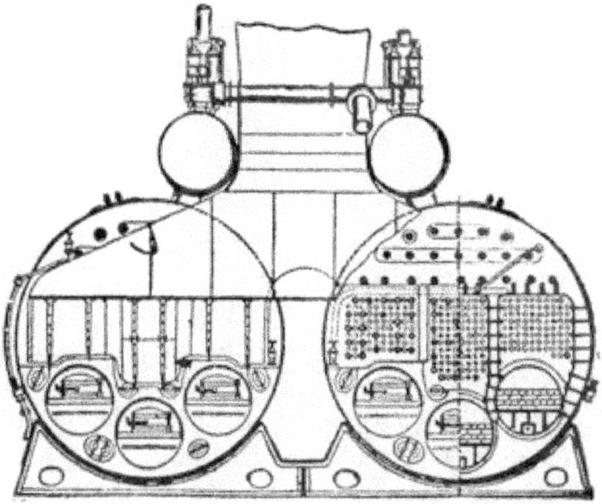

Fig 22.

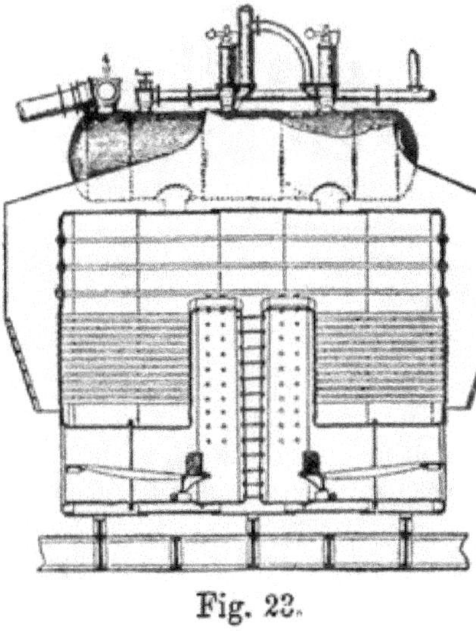

Fig. 23.

THE MARINE STEAM BOILER.

It will be seen from these drawings that there are three internal cylindrical furnaces at each end of these boilers, making in all six furnaces per boiler. The firing takes place at both ends. The flame and hot gases from each furnace, after passing over the bridge wall enter a flat-sided rectangular combustion chamber and then travel through tubes to the front uptake (*i. e.* the smoke bonnet or breaching), and so on to the chimney.

The sides of the combustion chambers are stayed to each other and to the shell plate of the boiler; the tops are strengthened in the same manner as the crowns of locomotive boilers, and the flat plates of the boiler shell are stayed together by means of long bolts, which can be lengthened up by means of nuts at their ends. Access is gained to the uptakes for purposes of cleaning, repairs of tubes, etc., by means of their doors on their fronts just above the furnace doors. The steam is collected in the large cylindrical receivers shown above each boiler. The material of construction is mild steel.

The following are the principal dimensions and other particulars of one of these boilers:

Length from front to back 20 feet.
Diameter of shell, 15 feet 6 inches.
Length of furnace, 6 feet 10 inches.
Diameter of furnace, 3 feet 10 inches.
Length of tubes, 6 feet 9 inches.
Diameter of tubes, 3½ inches.
No. of tubes, 516.
Thickness of shell plates, $\frac{15}{16}$.
Thickness of tube plates, $\frac{3}{4}$.
Grate area, 126½ square feet.
Heating surface, 4015 square feet.
Steam pressure, 80 lbs. per sq. inch.

Fig. 24 is a sketch of a modern marine boiler, which is only fired from one end, and is in consequence much shorter in proportion to its diameter than the type illustrated in figs. 22 and 23.

THE MARINE STEAM BOILER.

Marine boilers over nine feet in diameter have generally two furnaces, those over 13 to 14 feet, three, while the very largest boilers used on first-class mail steamers, and which often exceed fifteen feet in diameter, have four furnaces.

In the marine boiler the course taken by the products of combustion is as follows; the coal enters through the furnace doors on to the fire-bars, the heat and flames pass over the fire bridge into the flame or combustion chamber, thence through the tubes into the smoke-box, up the up-take and funnel into the air.

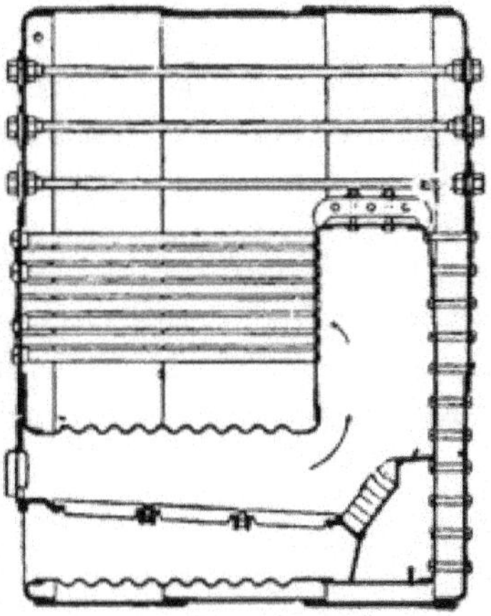

Fig. 24.

The fittings to a marine boiler are—funnel and air casings, uptakes and air casings, smoke boxes and doors, fire doors, bars, bridges, and bearers, main steam stop valve, donkey valve, safety valves and drain pipes, main and donkey feed check valves, blow-off and scum cocks, water gauge glasses on front and back of boiler, test water cock for trying density of water, steam cock for whistle, and another for winches on deck.

A fitting, called a blast pipe, is sometimes placed in the throat of the funnel. It consists of a wrought iron pipe, having a conical nozzle within the funnel pointing upwards, the other end being connected to a cock, which latter is bolted on to the steam space or dome of the boiler. It is used for increasing the intensity of the draft, the upward current of steam forcing the air out of the funnel at a great velocity; and the air having to be replaced by a fresh supply through the ash-pits and bars of the furnaces, a greater speed of combustion is obtained than would otherwise be due to simple draft alone.

THE MARINE STEAM BOILER.

Boilers are fitted with dry and wet uptakes, which differ from each other as follows:—The dry uptake is wholly outside the boiler, and consists of an external casing bolted on to the firing end of the boiler, covering the tubes and forming the smoke-box, and is fitted with suitable tube doors. A wet uptake is carried back from the firing ends of the boiler into its steam space, and is wholly surrounded by water and steam. The dry uptake seldom requires serious repair; but the wet uptake, owing to its exposure to pressure, steam, and water, requires constant attention and repair, and is very liable to corrosion, being constantly wetted and dried in the neighborhood of the water-line. The narrow water space between both front uptakes is also very liable to become burnt, owing to accumulation of salt. The flue boilers of many tugs and ferry boats are fitted with wet uptakes.

A superheater is a vessel usually placed in the uptake, or at the base of the funnel of a marine boiler, and so arranged that the waste heat from the furnaces shall pass around and through it prior to escaping up the chimney. It is used for drying or heating the steam from the main boiler before it enters the steam pipes to the engine. The simplest form of superheater consists of a wrought iron drum filled with tubes. The heat or flame passes through the tubes and around the shell of the drum, the steam being inside the drum. Superheaters are usually fitted with a stop valve in connection with the boiler, by means of which it can be shut off; and also one to the steam pipe of the engine; arrangements are also usually made for mixing the steam or working independently of the superheater.

A safety valve is also fitted and a guage glass; the latter is to show whether the superheater is clear of water, as priming will sometimes fill it up.

The special fittings of the marine boiler will be more particularly described hereafter as well as the stays, riveting, strength, etc., etc.

THE MARINE BOILER.

The use of the surface condenser in connection with the marine boiler was an immense step toward increasing its efficiency. In 1840 the average pressure used in marine boilers was only 7 or 8 lbs. to the square inch, the steam being made with the two-flue pattern of boiler, sea water being used for feed; as the steam pressure increased as now to 150 to 200 lbs. to the square inch, greater and greater difficulty was experienced from salt incrustation—in many cases the tubes did not last long and frequently gave much trouble, until the introduction of the surface condenser, which supplied fresh water to the boilers.

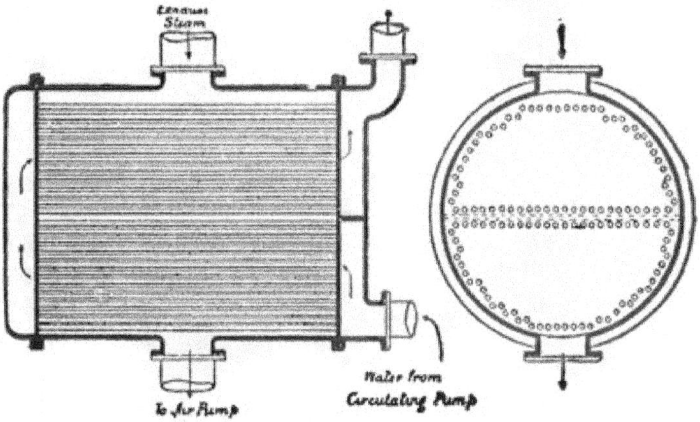

Fig. 25.

THE SURFACE CONDENSER.

The condenser is an oblong or circular box of cast iron fitted in one of two ways, either with the tubes horizontal or vertical: at each end are fixed the tube plates, generally made of brass, and the tubes pass through the plates as well as through a supporting plate in the middle of the condenser. Each end of the condenser is fitted with doors for the purpose of enabling the tube ends to be examined, drawn, or packed, as may be necessary. The tube ends are packed in various ways, and the tubes are made of brass, so as to resist the action of the water. The water is generally sucked through the tubes by the circulating pump, and the steam is condensed by coming in contact with the external surface of the tubes. In some cases the water is applied to the external surface, and the steam exhausted through the tubes; but this practice is now generally given up in modern surface condensers. The packing round the tube ends keeps them quite tight, and in the event of a split tube, a wooden plug is put in each end until an opportunity offers for drawing it and replacing with a new one.

The condenser may be made of any convenient shape. It sometimes forms part of the casting supporting the cylinders of vertical engines; it is also frequently made cylindrical with flat ends, as in fig 25. The ends form the tube plates to which the tubes are secured. The tubes are, of course, open at the ends, and a space is left between the tube plate and the outer covers, shown at each end of the condenser, to allow of the circulation of water as shown by the arrows.

Operation of the Condenser.

The cold water, which is forced through by a circulating pump, enters at the bottom, and is compelled to pass forward through the lower set of tubes by a horizontal dividing plate; it then returns through the upper rows of tubes and passes out at the overflow; the tubes are thus maintained at a low temperature.

The tubes are made to pass right through the condensing chamber, and so as to have no connection with its internal space. The steam is passed into the condenser and there comes in contact with the cold external surface of the tube, and is condensed, and removed, as before, by the air pump, as may be readily seen in the illustration (p. 65).

The advantages gained by the use of the surface condenser are: 1. The feed water is hotter and fresh; being hotter, it saves the fuel that would be used to bring it up to this heat; and being fresh, it boils at a lower temperature. 2. Not forming so much scale inside the boiler, the heat passes through to the water more readily; and as the scum cock is not used so freely, all the heat that would have been blown off is saved. Its disadvantages are that being fresh water and forming no scale on the boiler, it causes the boiler to rust.

THE MARINE BOILER.

It is often said that one engineer will get more out of a ship than another. In general it will be found that the most successful engineer is the man who manages his stokers best. It is very difficult on paper to define what is meant. It is a thing to be felt or seen, not described. * * * * The engineer who really knows his business will give his fires a fair chance to get away. He will work his engines up by degrees and run a little slowly for the first few moments.

WATER TUBE STEAM BOILERS.

A popular form of steam boiler in use in the United States and Europe is what is called the water tube boiler. This term is applied to a class of boiler in which the water is contained in a series of tubes, of comparatively small diameter, which communicate with each other and with a common

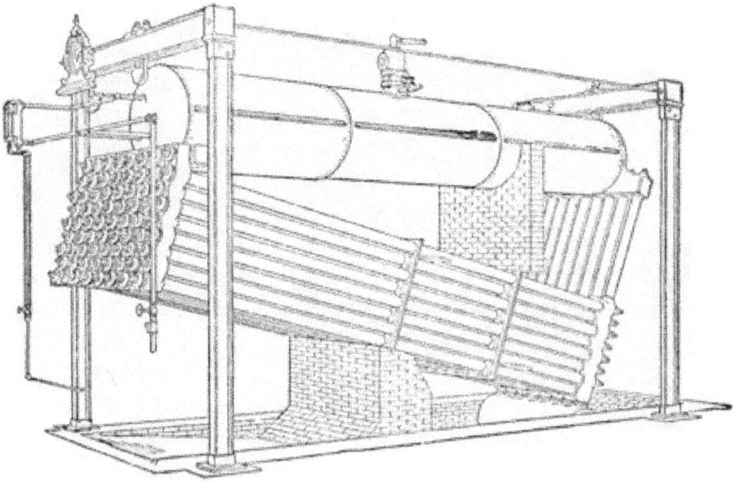

WATER TUBE BOILER.—Fig. 26.

steam-chamber. The flames and hot gases circulate between the tubes and are usually guided by partitions so as to act equally on all portions of the tubes. There are many varieties of this type of boiler of which the cut illustrates one: in this each tube is secured at either end into a square cast-iron head, and each of these heads has two openings, one communicating with the tube below and the other with the tube above; the

WATER TUBE STEAM BOILERS.

communication is effected by means of hollow cast-iron caps shown at the end of the tubes ; the caps have openings in them corresponding with the openings in the tube heads to which they are bolted.

In the best forms of the water tube boilers, it is suspended entirely independent of the brick work from wrought iron girders resting on iron columns. This avoids any straining of the boiler from unequal expansion between it and its enclosing walls and permits the brick work to be repaired or removed, if necessary, without in any way disturbing the boiler This design is shown in Fig. 26.

The distinguishing difference, which marks the water tube boiler from others,consists in the fact that in the former the small tubes are filled with water instead of the products of combustions ; hence the comparison, frequently made, between water-tube and *fire tube* boilers—the difference has been expressed in another way, " Water-tube vs. shell boilers," but the principle of steam production in both systems remains the same ; the heat from the combustible is transferred to the water through the medium of iron plates and in both, the furnaces, steam appliances, application of the draught, etc., is substantially the same. In another important point do the systems agree, *i. e.*, in the average number of pounds of water evaporated per lb. of combustible ; it is in the thoroughness of construction and skillfulness of adaptation to surroundings that produce the best results. Water tube or sectional boilers, have been made since the days of James Watt, in 1766, in many different forms and under various names. Owing, however, to the imperfection of manufacture the system, as compared to shell boilers, has been a failure until very recently ; various patterns of water-tube boilers are now in most favorable and satisfactory use. The advantages claimed for this form of steam generator are as follows :

1. Safety from disastrous explosions, arising from the division of the contents into small portions, and especially from details of construction which make it tolerably certain that the rupture will be local instead of a general violent explosion which liberates at once large masses of steam and water.

WATER TUBE STEAM BOILERS.

2. The small diameter of the tubes of which they are composed render them much stronger than ordinary boilers.

3. They can be cheaply built and easily repaired, as duplicate pieces can be kept on hand. The various parts of a boiler can be transported without great expense, trouble or delay; the form and proportions of a boiler can be suited to any available space; and, again, the power can be increased by simply adding more rows of tubes and increasing the grate area.

4. Their evaporative efficiency can be made equal to that of other boilers, and, in fact, for equal proportions of heating and grate surfaces, it is often a trifle higher.

5. Thin heating surface in the furnace, avoiding the thick plates necessarily used in ordinary boilers which not only hinder the transmission of heat to the water, but admit of overheating.

6. Joints removed from the fire. The use of lap welded water tubes with their joints removed from the fire also avoid the unequal expansion of riveted joints consequent upon their double thickness.

7. Quick steaming.

8. Accessibility for cleaning.

9. Ease of handling and erecting.

10. Economy and speediness of repairs.

The known disadvantages of these boilers are ·

1. They generally occupy more space and require more masonry than ordinary boilers.

2. On account of the small quantity of water which they contain, sudden fluctuations of pressure are caused by any irregularities in supplying the feed-water or in handling the fires, and the rapid and at times violent generation of steam causes it to accumulate in the contracted water-chambers, and leads to priming, with a consequent loss of water, and to overheated tubes.

WATER TUBE STEAM BOILERS.

3. The horizontal or inclined water tubes which mainly compose these boilers, do not afford a ready outlet for the steam generated in them. The steam bubbles cannot follow their natural tendency and rise directly, but are generally obliged by friction to traverse the tube slowly, and at times the accumulation of steam at the heated surfaces causes the tubes to be split or burned.

4. The use of water which forms deposits of solid matter still further increases the liability to overheating of the tubes. It has been claimed by some inventors that the rapid circulation of the water through the tubes would prevent any deposit of scale or sediment in them, but experience has proved this to be a grave error. Others have said that the expansion of the tube would detach the scale as fast as it was deposited and prevent any dangerous accumulation, but this also has been proved an error. Again, the use of cast iron about these boilers has frequently been a constant source of trouble from cracks, etc.

CARE OF WATER TUBE BOILERS.

The soot and ashes collect on *the exterior* of the tubes in this form of boilers, instead of inside the tubes, as in the tubular, and they must be as carefully removed in one case as in the other; this can be done by the use of blowing pipe and hose through openings left in the brick work; in using bituminous coal the soot should be brushed off when steam is down.

All the inside and outside surfaces should be kept clean to avoid waste of fuel; to aid in this service the best forms are provided with extra facilities for cleaning. For inspection, remove the hand holes at both ends of the tubes, and by holding a lamp at one end and looking in at the other the condition of the surface can be freely seen. Push the scraper through the tube to remove sediment, or if the scale is hard, use the chipping scraper made for that purpose.

Hand holes should be frequently removed and surfaces examined, particularly in case of a new boiler. In replacing

CARE OF WATER TUBE BOILERS.

hand hole caps, clean the surfaces without scratching or bruising, smear with oil and screw up tight.

The mud drum should be periodically examined and the sediment removed; blow-off cocks and check valves should be examined each time the boiler is cleaned; when surface blow-cocks are used they should be often opened for a few minutes at a time; be sure that all openings for air to boiler or flues *except through the fire,* are carefully stopped.

If a boiler is not required for some time, empty and dry it thoroughly. If this is impracticable, fill it quite full of water and put in a quantity of washing soda; and external parts exposed to dampness should receive a coating of linseed oil. Avoid all dampness in seatings or coverings and see that no water comes in contact with the boiler from any cause.

Although this form of boiler is not liable to destructive explosion, the same care should be exercised to avoid possible damage to boilers and expensive delays.

SECTIONAL BOILERS.

Probably one of the first sectional boilers brought into practical use is one made of hollow cast iron spheres, each 8 inches in diameter, externally, and $\frac{3}{8}$ of an inch thick, connected by curved necks $3\frac{1}{4}$ inch in diameter. These spheres are held together by wrought iron bolts and caps, and in one direction are cast in sets of 2 or 4, which are afterwards drawn together so as to give more or less heating surface to the boiler according to the number used.

NOTE.

Owing to their multiplication of parts all sectional, including water tube boilers, should be made with unusual care in their details of construction, setting, fittings and proportions. It is to the attention paid to these "points" that the sectional boilers are now coming into more general favor.

LOCOMOTIVE BOILERS.

The essential features of locomotive boilers are dictated by the duties which they have to perform under peculiar conditions. The size and the weight are limited by the fact that the boiler has to be transported rapidly from place to place, and also that it has to fit in between the frames of the locomotive; while at the same time, the pressure of the steam has to be very great in order that with comparatively small cylinder the engine may develop great power; moreover, the quantity of water which has to be evaporated in a given time is very considerable. To fulfil these latter conditions a large quantity of coal must be burned on a fire grate of limited area; hence intense combustion is necessary under a forced blast. To utilize advantageously the heat thus generated, a large heating surface must be provided and this can only be obtained by passing the products of combustion through a great number of tubes of small diameter.

The forced draught in a locomotive boiler is obtained by causing the steam from the cylinders, after it has done its work to be discharged into the chimney by means of a pipe called the blast pipe; the lower portion of this consists of two branches, one in communication with the exhaust port of each cylinder. As each puff of steam from the blast pipe escapes up the chimney it forces the air out in front of it, causing a partial vacuum, which can only be supplied by the air rushing through the furnace and tubes.

The greater the body of steam escaping at each puff, and the more rapid the succession of puffs, the more violent is the action of the blast pipe in producing a draught, and consequently this contrivance regulates the consumption of fuel and the evaporation of water to a certain extent automatically, because when the engine is working its hardest and using the most steam, the blast is at the same time most efficacious.

The blast pipe is, perhaps, the most distinctive feature of the locomotive boiler, and the one which has alone rendered it possible to obtain large quantities of steam from so small a

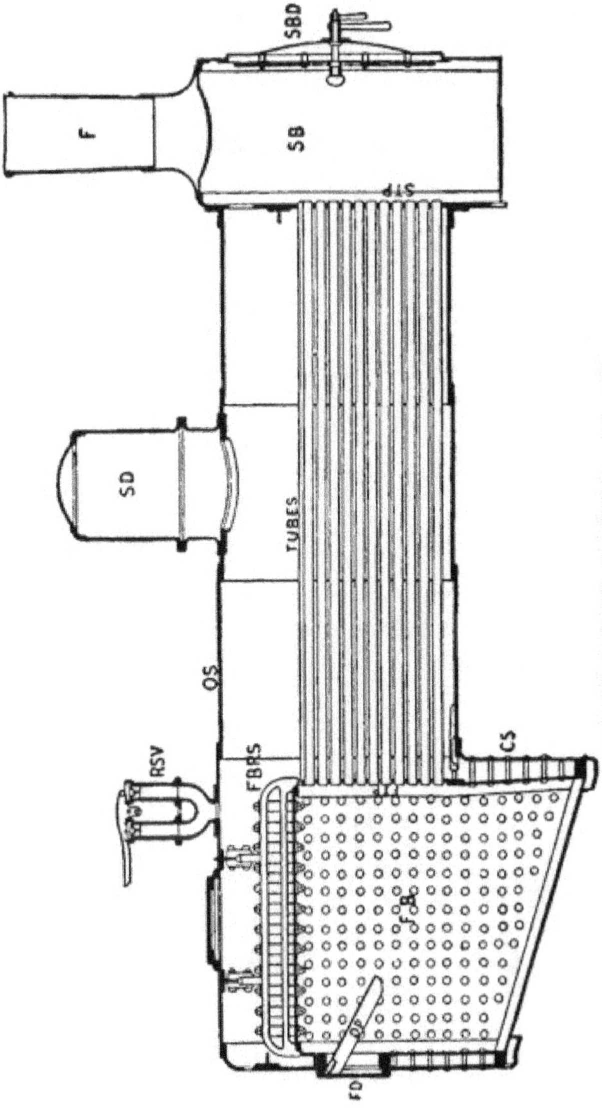

LOCOMOTIVE BOILER.—Fig. 27.

THE LOCOMOTIVE BOILER.

generator. The steam blast of the locomotive has been compared to the breathing apparatus of a man, and has rendered the mechanism described nearer a live thing than any other device man has ever produced.

On account of the oscillations, or violent motions to which the boiler of locomotive engines are subject, weighted safety valves are not possible to be used and springs are used instead to hold the valves in place.

The locomotive form of steam boiler is sometimes used for stationary engines, but owing to extra cost and increased liability to corrode in the smaller passage they are not favorites.

DESCRIPTION OF PAGE ILLUSTRATION.

In fig. 27, F B represents the fire box or furnace; F D, fire door: D P, deflector plate; F T P, fire box tube plate; F B R S, fire box roof stays; S T P, smoke box tube plate; S B, smoke box; S B D, smoke box door; S D, steam dome; O S, outer shell; R S V, Ramsbottom safety valve; F, funnel or chimney.

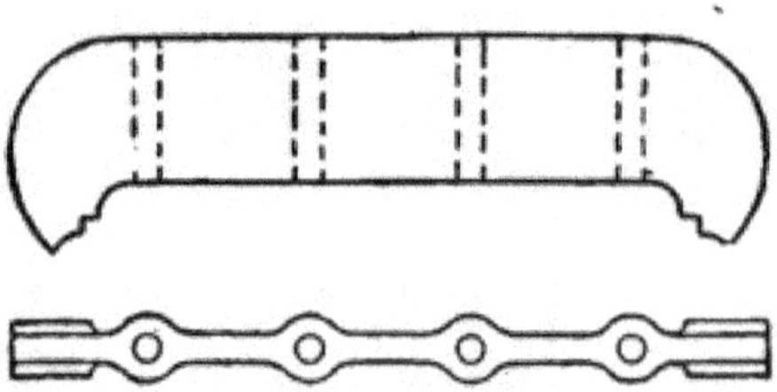

Fig. 28.

The crown plate of the fire-box being flat requires to be efficiently stayed, and for this purpose girder stays called fox roof stays are mostly used, as shown in the figure. The stays are now made of cast steel for locomotives. They rest at the two ends on the vertical plates of the fire-box, and sustain the pressure on the fire-box crown by a series of bolts passing through the plate and girder stay, secured by nuts and washers. Fig. 28 is a plan and elevation of a wrought-iron roof stay.

Another method adopted in locomotive types of marine boilers for staying the flat crown of the fire-box to the circular upper plate is shown in fig. 29—namely, by wrought-iron vertical bar stays secured by nuts and washers to the fire-box with a fork end and pin to angle-iron pieces riveted to the boiler shell.

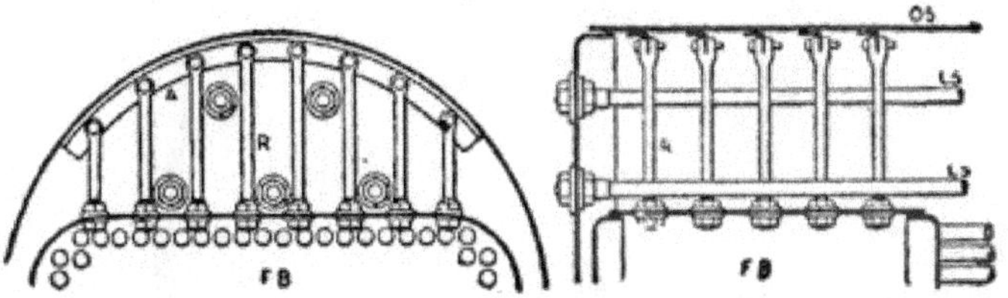

Fig. 29.

The letters in this figure refer to the same parts of the boiler as do those in fig. 27, *i.e.*, F B to the fire-box, etc., etc.

It was formerly the custom to make the tubes much longer than shown in the fig., with the object of gaining heating surface; but modern experience has shown that the last three or four feet next the smoke box were of little or no use, because, by the time the products of combustion reached this part of the heating surface, their temperature was so reduced that but little additional heat could be abstracted from them. The tubes, in addition to acting as flues and heating surface, fulfil also the function of stays to the flat end of the barrel of the boiler, and the portion of the fire box opposite to it.

In addition to the staying power derived from the tubes, the smoke box, tube plate and the front shell plate are stayed together by several long rods.

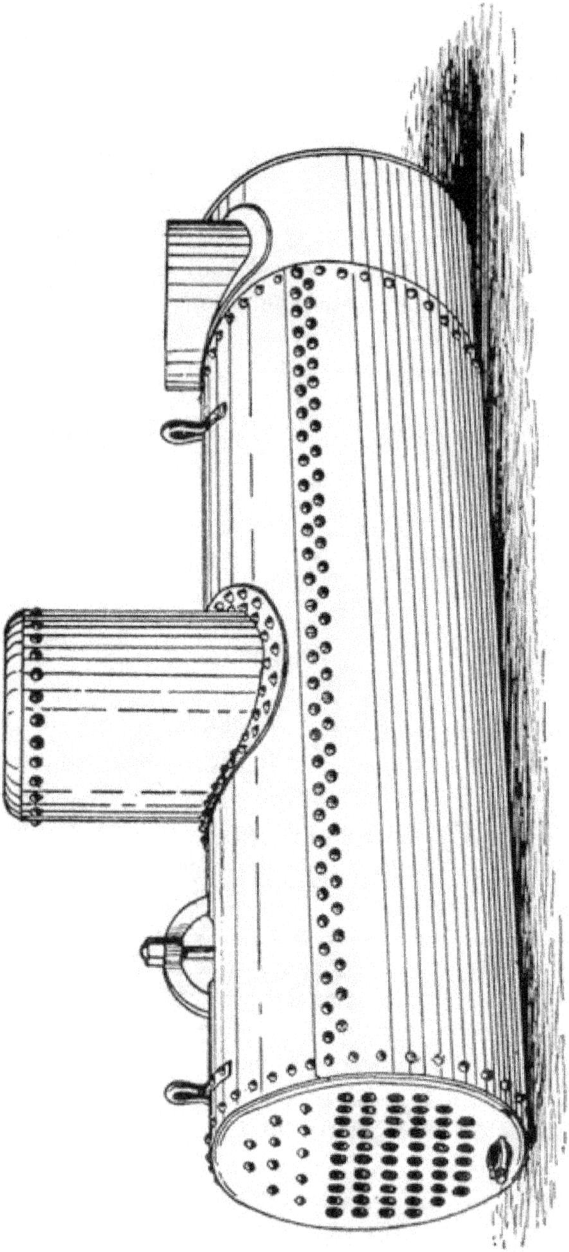

THE HORIZONTAL TABULAR BOILER.—FIG. 30.

STANDARD HORIZONTAL TUBULAR STEAM BOILER.

TABLE OF SIZES, PROPORTIONS, ETC.:

Diameter of Shell.	Length of Shell.	Gauge of Shell.	Gauge of Heads.	Number of Tubes.	Diameter of Tubes.	Length of Tubes.	Square feet of Heating Surface.	Nominal Horse Power.
72 in.	19 ft. 4 in.	3-8 in.	1-2 in.	80	4 in.	18 ft. 0 in.	1,500	100
72 "	18 " 4 "	3-8 "	1-2 "	86	3½ "	17 " 0 "	1,500	100
72 "	17 " 4 "	3-8 "	1-2 "	108	3 "	16 " 0 "	1,500	100
66	8 " 4 "	3-8 "	1-2 "	74	3½ "	17 " 0 "	1,350	90
66 "	17 " 4 "	3-8 "	1-2 "	92	3 "	16 " 0 "	1,350	90
60 "	18 " 3 "	3-8 "	1-2 "	78	3 "	17 " 0 "	1,200	80
60 "	17 " 3 "	3-8 "	1-2 "	76	3 "	16 " 0 "	1,125	75
60 "	16 " 3 "	3-8 "	1-2 "	77	3 "	15 " 0 "	1,050	70
60 "	16 " 3 "	3-8 "	1-2 "	70	3 "	15 " 0 "	975	65
60 "	16 " 3 "	3-8 "	1-2 "	64	3 "	15 " 0 "	900	60
54 "	17 " 3 "	5-16 "	7-16 "	60	3 "	16 " 0 "	900	50
54 "	17 " 3 "	5-16 "	7-16 "	56	3 "	16 " 0 "	825	55
54 "	16 " 3 "	5-16 "	7-16 "	52	3 "	15 " 0 "	750	50
54 "	16 " 3 "	5-16 "	7-16 "	46	3 "	15 " 0 "	675	45
54 "	16 " 3 "	5-16 "	7-16 "	40	3 "	15 " 0 "	600	40
48 "	17 " 2 "	5-16 "	7-16 "	50	3 "	16 " 0 "	750	50
48 "	16 " 2 "	5-16 "	7-16 "	48	3 "	15 " 0 "	675	45
48 "	16 " 2 "	5-16 "	7-16 "	42	3 "	15 " 0 "	600	40
42 "	16 " 2 "	1-4 "	3-8 "	36	3 "	15 " 0 "	525	35
42 "	15 " 2 "	1-4 "	3-8 "	32	3 "	14 " 0 "	450	30
42 "	14 " 2 "	1-4 "	3-8 "	28	3 "	13 " 0 "	375	25
36 "	14 " 2 "	1-4 "	3-8 "	36	2½ "	13 " 0 "	375	25
36 "	14 " 2 "	1-4 "	3-8 "	28	2½ "	13 " 0 "	300	20
36 "	13 " 2 "	1-4 "	3-8 "	20	2½ "	12 " 0 "	225	15
36 "	12 " 2 "	1-4 "	3-8 "	14	2½ "	11 " 0 "	150	10

NOTE.

In estimating the horse power by means of the above table, 15 square feet has been allowed for each horse power, and the number of feet in each boiler is given *in round numbers.* This table is one used in every day practice by boiler makers.

THE FLUE BOILER.

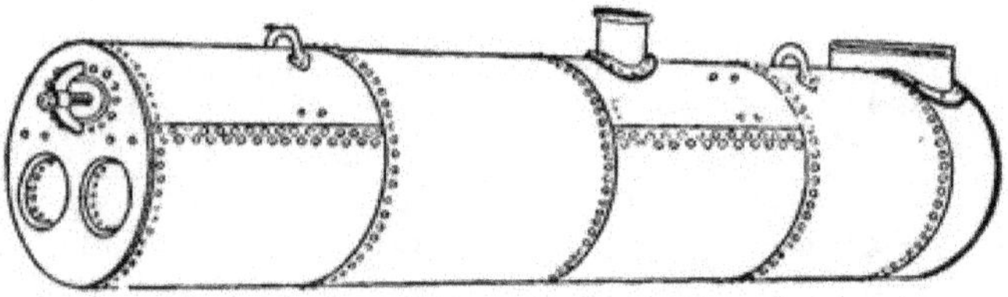

THE TWO FLUE BOILER.—Fig. 31.

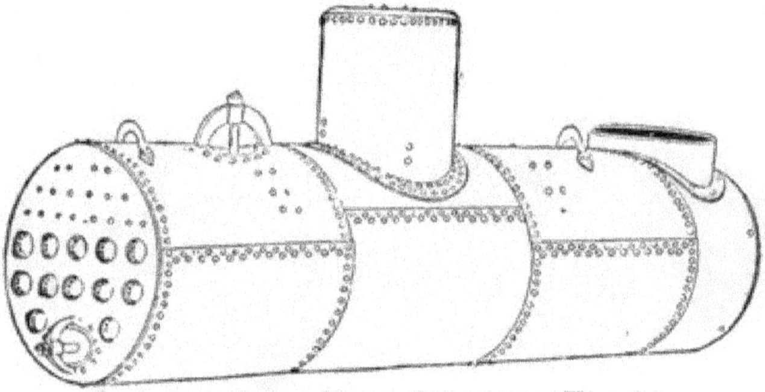

THE SIX INCH FLUE BOILER.—Fig. 32.

THE HORIZONTAL TUBULAR STEAM BOILER.

The great majority of stationary boilers are cylindrical or round shaped, because

1. The cylindrical form is the strongest.

2. It is the cheapest.

3 It permits the use of th...er metal.

4. It is the safest.

5. It is inspected without difficulty.

6. It is most symmetrical.

7. It is manufactured easier.

8. It resists internal strain better.

9. It resists external strain also.

10. It can be stayed or strengthened better.

11. It encloses the greatest volume with least material.

12. It is the result of many years experience in boiler practice.

13. It is the form adopted or preferred by all experienced engineers.

It follows, too, that *the horizontal tubular boiler,* substantially as shown in Fig. 30, is the standard steam boiler ; engineers and steam power owners cling with great tenacity to this approved form, which is an outgrowth of one hundred years' experience in steam production.

In the plain horizontal tubular boiler shown in cuts, the shell is filled with as many small tubes varying from two inches to four inches in diameter as is consistent with the circulation and steam space. In firing this type of boiler the combustion first takes place under the shell, and the products, such as heat, flame, and gas, pass through the small tubes to the chimney, although in the triple draught pattern of the tubular boiler, the heat products pass, as will hereafter be explained, a second time through the boiler tubes, making three turns before the final loss of the extra heat takes place.

THE HORIZONTAL TUBULAR STEAM BOILER.

The illustrations on pages 78 and 80 exhibit the gradual advances to the horizontal tubular by the two-flued boiler (fig. 31) of the six flues (fig. 32) and of the locomotive Portable Boiler (fig. 33). The vertical or upright tubular boiler is but another modification of the horizontal tubular.

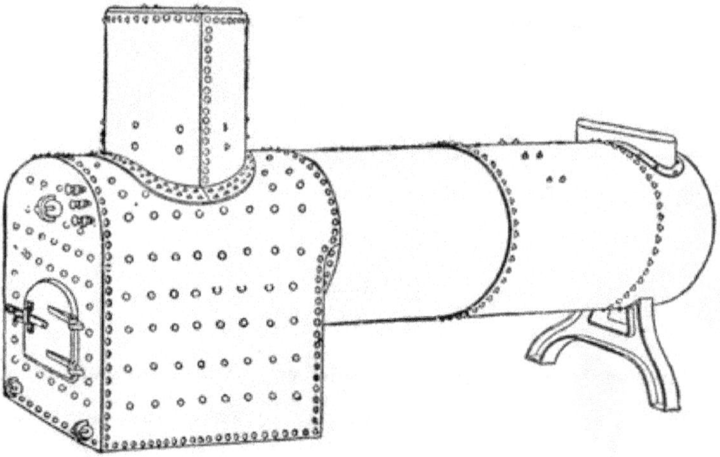

THE LOCOMOTIVE PORTABLE BOILER.—Fig. 33.

In parts of the vertical boiler there is very little circulation and the corrosion on the inner side is such as to wear the boiler rapidly. In the ash pit, ashes and any dampness that may be about the place also causes rapid corrosion. The upper part of the tubes and tube sheet are frequently injured; for instance, if the tubes pass all the way through to the upper tube sheet, providing there is no cone top, when the fire is first made under the boiler, combustion at times does not take place until the gases pass nearly through the tubes. The water usually being carried below the tube sheet there is a space left above the water line, where there is neither steam nor water, and the heat is so great that the ends of the tubes are burned and crystalized, and the tube sheet is often cracked and broken by this excessive heat before the steam is generated The first difficulty is experienced in "the legs" of the Portable Locomotive boiler—hence the general verdict of steam users in favor of the round shell, many-tubed boiler.

PARTS OF THE TUBULAR BOILER.

THE SHELL. This is the round or cylindrical structure which is commonly described as the boiler, in which are inserted the braces and tubes, and which sustains the internal strain of the pressure of the steam, the action of the water within, and the fire without.

THE DRUM. This part is sometimes called the dome, and consists of an upper chamber riveted to the top of the boiler for the purpose of affording more steam space.

THE TUBE SHEETS. These are the round, flat flanged sheets forming the two ends of the boiler, into which the tubes are fastened.

THE MANHOLE COVER. This is a plate and frame commonly opening inwards and large enough to admit a man into the interior of the boiler. These openings are sometimes made on the top and sometimes at the end of the boiler. Manhole openings in steam boilers should invariably be located in the head of the boiler, except in rare cases that may arise, when circumstances require it to be placed in the shell. The manhole, so placed, will not materially reduce the strength of the boiler, and from this position it can more readily be seen that the boiler is kept in proper condition. The proper sizes for manholes are 9x5 and 10x16, according to circumstances. These are amply large for general use and no material advantage is gained by increasing them.

THE HAND HOLE PLATES. These are similar arrangements to the manhole cover, except as to size. They are made large enough to admit the hand into the boilers for the purpose of removing sediment and they are also used for the purpose of inspecting the interior of the boiler. Two are usually put in each boiler, one front and one in the rear.

THE BLOW OFF. This consists of pipes and a cock communicating with the bottom of the boiler for the purpose of blowing off the boiler or of running off the water when the former needs cleaning.

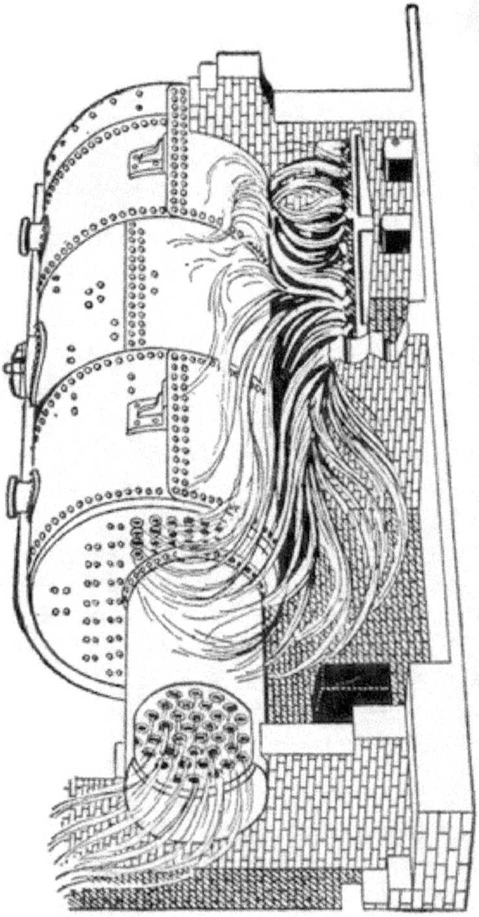

THE TRIPLE DRAUGHT TUBULAR BOILER.—Fig. 34.

THE TRIPLE DRAUGHT TUBULAR BOILER.

This boiler, which is extensively used by the manufacturers of New England, is, as will be seen by the illustration, of the horizontal tubular class, and is essentially different from the well-known type only in the arrangement of the tubes. The method secures the passage of the products of combustion through the same shell twice; forward through a part of the tubes, and backwards through the remaining ones. The manner of accomplishing this result can be best described by explaining how a common tubular boiler may be remodelled so as to carry out this principle.

A cylindrical shell, as shown in Fig. 34—of sufficient size to encircle about one-half of the tubes, is attached to *the outside of*

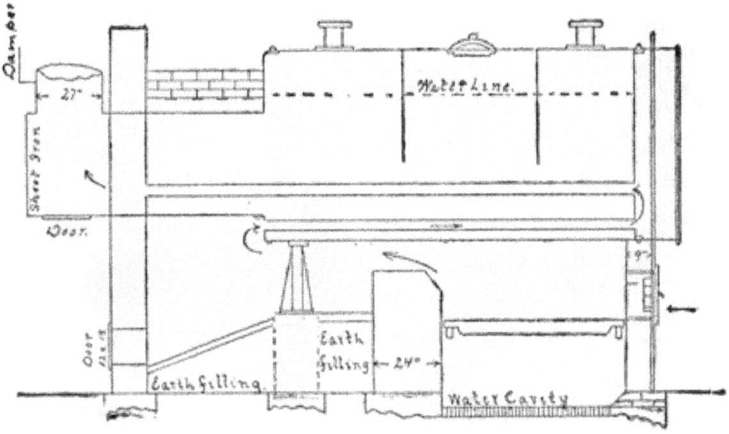

Fig. 35.

the rear head below the water line, and extended backward to the back end of the setting. The encircled tubes are lengthened and carried backward to the same point; the extension is closed in and made to communicate with the boiler proper; the inner tubes emerge to the flue leading to the chimney and the old connection from the smoke arch is cut off. With this arrangement, the outer tubes of the boiler—a cluster on each side of the supplementary shell carry the products of combustion forward to the front smoke arch, and the inner tubes carry them backward to the chimney.

THE TRIPLE DRAUGHT TUBULAR BOILER.

Fig. 35 exhibits the boiler in half section and shows the course of the heat products through *one* of the outer tubes and returning through the boiler by *one* of the inner cluster.

Fig. 36 (page 84) shows the boiler sectionally, over the bridge wall; the *shaded* tube ends exhibit the cluster which return the heat products to the rear of the boiler, after being brought forward by the two outer clusters which are left unshaded.

This arrangement of the tubes gives several advantages:

1. It enables an exceedingly high furnace temperature, without loss at the chimney.

2. By dividing the heat into these currents a more equal expansion and contraction is secured. This is an important point secured.

3. In this system the tubes are almost equally operative.

4. The extra body of water immediately over the furnace is both an element of safety and a reservoir of power.

5. The outlet for the waste products of combustion is found in this style of boiler in a more convenient position at *the rear end* of the boiler.

6. The boiler being self contained, can be used in places where height of story is limited.

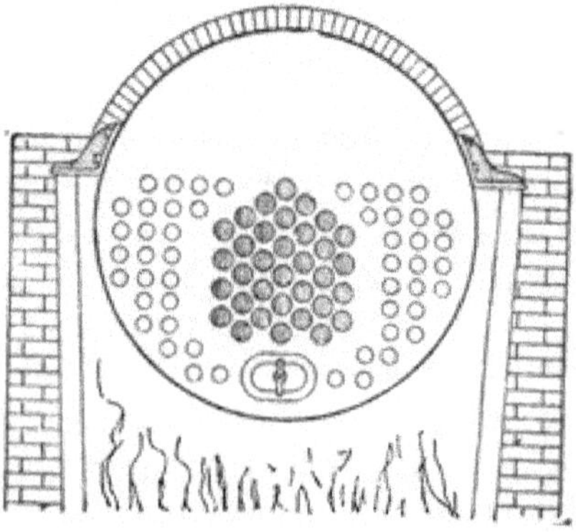

Fig. 36.

SPECIFICATION FOR 125 HORSE POWER BOILER.

For one Horizontal Tubular Boiler 72 inches diameter 18 feet long for..............................of........

Type.

The boiler to be of the Horizontal Tubular type with all castings and mountings complete.

Dimensions.

Boiler 72 inches diameter and 18 feet long. Each boiler to contain 90 best lap welded tubes 3½ inches diameter by 18 feet long, set in vertical and horizontal rows with a space between them vertically and horizontally of no less than one inch and one quarter (1¼) except central vertical space, which is to be three inches (3). No tube to be nearer than two and one-half inches (2½) to shell or boiler. Holes through heads to be neatly chamfered off. All tubes to be set by Dudgeon Expander and slightly flared at front end, turned over and beaded down at back end.

Quality and Thickness of Steel Plates.

Shell plates to be ½ inch thick of homogeneous steel of uniform quality having a tensile strength of not less than 65,000 lbs. Name of maker, brand and tensile strength to be plainly stamped on each plate.

Heads to be of same quality as plates of shell in all particulars ⅝ inch thick. Bottom of shell to be of one plate, and no plate to be less than 7 feet wide. Top of shell to be in three plates. All plates planed before rolling, and all joints fullered not caulked.

Flanges.

All flanges to be turned in a neat manner to an internal radius of not less than two inches (2) and to be clear of cracks, checks or flaws.

SPECIFICATION FOR STEAM BOILER.

Riveting.

Boilers to be riveted with ¾ inch rivet throughout. All girth seams to be double riveted. All horizontal seams to be double riveted. Rivet holes to be punched or drilled so as to come fair in construction. No drift pins to be used in construction of the boilers.

Braces.

All braces to be of the crowfoot pattern, one and one-eighth (1⅛) inch diameter and the shortest to be no less than four feet (4) long and of sufficient number for thorough bracing, and to bear uniform tension.

Manholes, Hand Holes and Thimbles.

One manhole in top of each boiler with heavy cast iron frame riveted on middle of centre plate ; one manhole near the bottom of each front head ; head reinforced with a wrought iron ring two inches (2) square, riveted to heads with flush countersunk rivets two inches (2) pitch and to have all the necessary bolts, plates, guards and gaskets ; two six inch thimbles riveted to top of each boiler, each to have a planed face ; one heavy 6 inch flange on bottom of each boiler, 12 inches from back end to centre of flange. There must be two braces, one on each side of manhole in front head ; also to have three braces opposite manhole on back head below tubes.

Lugs.

Four (4) lugs riveted on each side of boilers, of good and sufficient size, with six one-inch rivets in each lug.

Castings.

Each boiler to have a complete set of castings consisting of ornamental flush fronts containg tube, fire and ash-pit doors, and provide the best stationary grate bars as may be selected by buyer, with the necessary fixtures, all bearing bars, britching plates, dead plates, binder bars, back cleaning out doors with frames. Anchor bolts and buck stays. The fire door to contain adjustable air opening and to be protected with fire shields. One heavy cast iron arch over each boiler.

SPECIFICATION FOR STEAM BOILER.

Testing.

Boilers to be tested with a water pressure of 200 lbs. per square inch and certificate of such test having been made shall be furnished with boiler. Test of boiler to be under direction of such steam boiler Insurance Company as may be selected by buyer.

Quality and Workmanship.

All boilers tc be made in the best workmanlike manner and all material of their respective kinds to be of the best, and in strict accordance with specification.

Fittings and Mountings.

The boiler to be furnished with the following: One four inch heavy mounted safety valve. One six inch flanged globe valve. Two two inch best globe valves. Two two inch check valves. One eight inch dial nickle plated steam gauge. One low water alarm gauge. One set of fire irons for two boilers consisting of hoe, poker, slice bar and shovel.

Drawings.

All drawings furnished for masons in setting the boilers.

Duty of Boiler.

The boiler to develop 120 horse power and to work under a constant pressure varying from 125 to 150 lbs. to the square inch.

All rivets are to be 2½ and 1½ inch pitch. The pitch line of the rivets to be not nearer 1½ inches to the edge of the sheet.

To be 8 lug plates for each boiler not less than 2 feet long, 8 inches wide, and one inch thick.

There shall be six 1 inch anchor rods running front to rear of each boiler, in the brick work.

These boilers and all their fronts, fittings and connections will be subject to the inspection of........................,

MARKS ON BOILER PLATES.

Something has been said under another heading of the nature and requisite quality of the materials entering into the structure of the boiler. Too much emphasis cannot be laid upon the necessity for the use of the very best iron and steel that can be manufactured, and the most skillful and thorough workmanship that can be performed in constructing the boiler.

It is becoming the practice, both for land and marine boilers, for boiler plate makers to furnish "test pieces" from each sheet or plate that goes into the construction of a boiler, and a sheet showing the tensile strength of each sheet or plate that enters into its make up.

But irrespective of this practice each plate entering into boiler construction will be found to have one of the following marks, which designate its quality and method of manufacture. The name "Charcoal Iron" is used because in its manufacture wood charcoal is employed instead of mineral fuel.

"Charcoal No. 1 Iron" (C. No. 1) is made entirely of charcoal iron. It has a tenacity of 40,000 pounds per square inch in the direction of the fibre. It is hard, but not very ductile, and should never be used for flanging.

"Charcoal Hammered No. 1 Shell Iron" (C. H. No. 1 S.), although not necessarily hammered, has been worked up before it is rolled into plates. It has a tenacity of 50,000 to 55,000 pounds per square inch in the direction of the fibre. It is rather hard iron, and should not be flanged. It is used for the outside shell of boilers.

"Flange iron" (C. H. No. 1 F.), is a ductile material which can be flanged in every direction. It has a tenacity of 50,000 to 55,000 pounds per square inch along the fibre.

"Fire Box Iron" (C. H. No. 1 F. B.), is a harder quality, designed especially to withstand the destructive effect of the impinging flame, and is used for boxes and flue-sheets.

The letters in the brackets exhibit the plate stamp.

Cast iron and copper were used in an early day for steam boilers and the former is still extensively used for certain forms of low pressure steam heaters made for various purposes, such as green-houses, &c.

CONSTRUCTION OF BOILERS.

In selecting a boiler, the most efficient design will be found to be that in which *the greatest amount of shell surface is exposed to direct heat.* It is the direct heating surface that does the bulk of the work and every tendency to reduce it, either in the construction or setting of the boiler, should be avoided. The smaller the amount of surface enclosed by or in contact with the setting, the better results will be obtained.

A boiler with a bad circulation is the bane of an engineer's existence. Proper circulation facilities constitute one of the chief factors in the construction of a successful and economical boiler. In tubular boilers the best practice places the tubes in vertical rows, leaving out what would be the centre row. The circulation is up the sides of the boiler and down the centre. Tubes set zig-zag to break spaces impede the circulation and are not considered productive of the best results.

The surface from which evaporation takes place should be made greater as the steam pressure is reduced, that is to say, as the size of the bubbles of steam become greater. To produce 100 lbs. of steam per hour at atmospheric pressure this surface should not be less than 732 square feet, which may be reduced to 146 square feet for steam at 75 lbs. pressure, and to 73 feet for steam at a pressure of 150 lbs. It is for this reason that triple-expansion engines can be worked with smaller boilers than are required with engines using steam of lower pressure. The amount of steam space to be permitted depends upon the volume of the cylinders and the number of revolutions made per minute. For ordinary engines it may be made a hundred times as great as the average volume of steam generated per second.

A volume of heated water in a boiler performs the same office in furnishing a steady supply of steam, as a fly-wheel does to an engine in insuring uniformity of speed; hence the centre space of a boiler should be ample, in order to take advantage of this reserve force.

QUALITY OF STEEL PLATES.

Steel for boilers is always of the kind known as low steel, or soft steel, and is, properly speaking, *ingot iron*, all of its characteristics being those of a tenacious, bending, equal grained iron, and quite different from true steels, such as knife blades, cutting tools, etc., are composed of. Steel is rapidly displacing iron in boiler construction, as it has greater strength for the same thickness, than iron ; and, except in rare instances, where the nature of the water available for feed renders steel undesirable, iron should not be used for making boilers, careful tests having shown it to be vastly inferior to steel in many important features.

Good boiler steel up to one-half inch in thickness should be capable of being doubled over and hammered down on itself without showing any signs of fracture, and above that thickne s it should be capable of being bent around a mandrel of a diameter equal to one and one-half times the thickness of the plate, to an angle of 180 degrees without sign of distress. Such bending pieces should not be less in length than sixteen times the thickness of the plate.

On this test piece the metal should show the following physical qualities :

Tensile strength, 55,000 to 65,000 pounds per square inch.

Elongation, 20 per cent. for plates three-eighths inch thick or less.

Elongation, 22 per cent. for plates from three-eighths to three-fourths inch thick.

Elongation, 25 per cent. for plates over three-fourths inch thick.

The cross sectional area of the test piece should be not less than one-half of one square inch, *i. e.*, if the piece is one-fourth inch thick, its width should be two inches ; if it be one-half inch thick, its width should be one inch. But for heavier material the width shall in no case be less than the thickness of the plate.

NICKEL STEEL BOILER PLATES.

It has been found that the addition of about three per cent. (3.16 to 3.32) of nickel to ordinary soft steel produces most favorable results ; thus it has been shown by Riley that a particular variety of nickel steel presents to the engineer *the means of nearly doubling boiler pressures without increasing weight or dimensions.*

In a recent experiment made with Bessemer steel rolled into three-fourths inch plates, from which a number of test specimens were cut, the elastic limit was respectively 59,000 pounds and 60,000 pounds. The ultimate tensile strength was 100,000 pounds and 102,000 pounds, respectively. The elongation was 15½ per cent. in each specimen, and the reduction of area at fracture was 29½ per cent. and 20½ per cent. respectively. These figures show that the elastic limit and ultimate tensile strength were raised by the nickel alloy to almost double the limits reached in the best grades of boiler plate steel, and the elongation was reduced to a scarcely appreciable extent.

The experiment had for its object, the reproduction, as nearly as possible, of the alloy used in the nickel steel armor plate made at Le Creusot, France, and the results were reported to the Secretary of the Navy at Washington. The new plate showed a percentage of 3.16 nickel, as against 3.32 for the imported plate.

RIVETING.

When the materials are of best quality, then there only remains to rivet and stay the boiler. *Riveting* is of two kinds, single and double. Fig. 37 shows the method of single riveting, and Figs. 38 and 39 show the plan and cross-section of double riveted sheets.

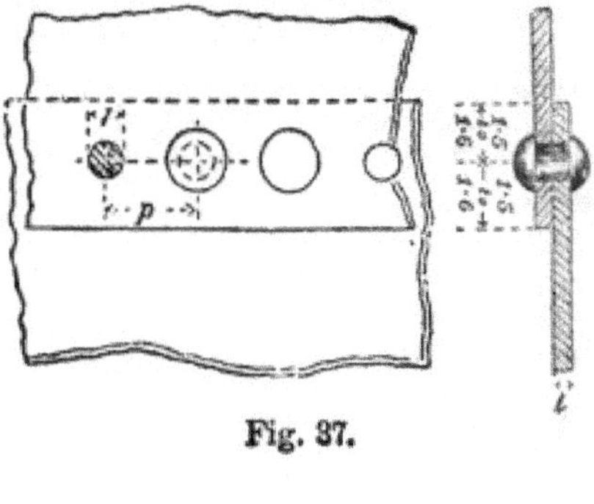

Fig. 37.

Double riveting consists in making the joints of boiler work with two rows of rivets instead of one—nearly always, horizontal seams are double riveted as well as domes where they join upon the boiler. Usually all girth

RIVETING.

seams,—those running round the body of the boiler, are single riveted. The size of the rivets is in proportion to the diameter of the boiler, being ⅝, ¾ and ⅞ as required in the specification.

Rivet holes are made by punching or drilling, according to the material in which they are made. In soft iron and mild steel they may safely be punched, but in metal at all brittle the holes should be drilled.

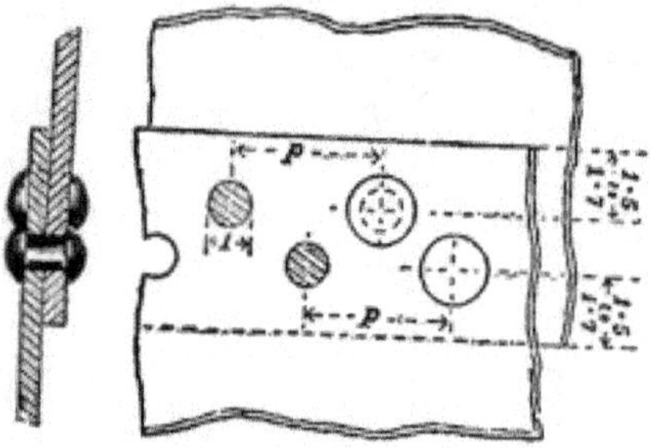

Fig. 38.

Rivets are driven by hand, by steam riveting machines or by an improved pneumatic machine which holds the sheet together and strikes a succession of light blows to form the head of the rivet while hot. Rivets are made both of iron and steel, and there are certain well-known brands of such excellent quality that they are almost exclusively used in boiler work.

A place where skill is shown in boiler construction is in laying out the rivet holes, with a templet so that the sheets come exactly together with the holes so neatly opposite that the dreaded drift pin does not have to be used.

In these figures the letters P and p refer to the "pitch of the rivets," *i. e.*, the part from centre to centre, and the dimensions given at the sides indicate the amount of lap given in inches and tenths of inches—the diameter of the rivet (1″) is also shown, and the turned over portion of the shank of the rivet is shown by dotted lines.

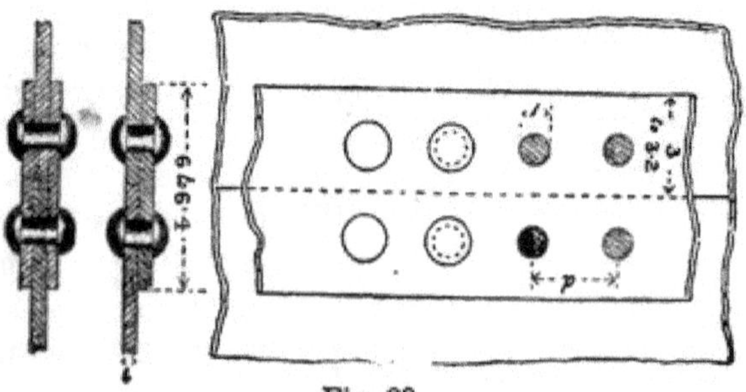

Fig. 39.

RIVETING.

No riveted boiler work can be considered fairly proportioned unless the strength of the plate between the rivets is fully equal to the strength of the rivets themselves. A margin (or net distance from outside of holes to edge of plate) equal to the diameter of the drilled hole has been found sufficient.

Rivets should be made of good charcoal iron or of a very soft mild steel, running between 50,000 and 60,000 pounds tensile strength and showing an elongation of not less than ninety per cent. in eight inches, and having the same chemical composition as specified for plates.

A long rivet, holding thick plates together, is rarely tight except immediately under the head. The heads are set and the centre cooled before the hole is properly filled. If it is a very long rivet there is a chance of the contraction fracturing the head of the rivet, In the Forth Bridge, which is made of very heavy plate girders, the rivets, first carefully fitted, were driven tight into the holes, the burr around the holes was removed and the ends of the rivets heated to a sufficient degree to enable them to be closed over.

A simple mathematical deduction shows that a circle seam has just one-half the strain to carry as a longitudinal seam, under the same pressure and with the same thickness of metal, hence the custom of single riveting the former and double riveting the latter, or longwise seams.

Different Modes of Riveting.

Chain Riveting.	Zig Zag Riveting.	Treble Riveting.	Unequal Pitches.
o o	o	o o	o o o
	o	o	
o o	o	o o	o o
	o	o	
o o	o	o o	o o o
	o	o	
o o	o	o o	o o
	o	o	
o o	o	o o	o o o

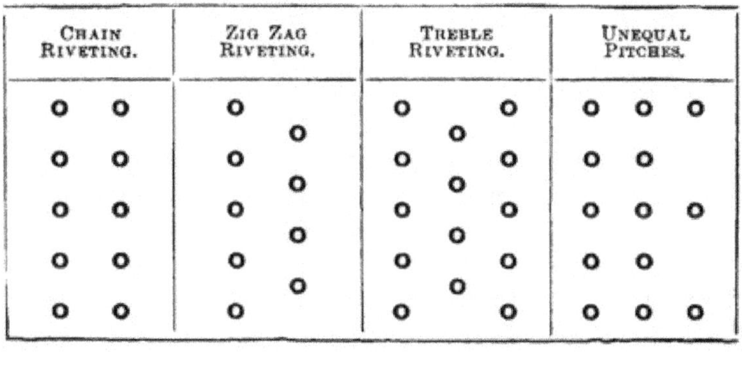

RIVETING.

In fig. 41 may be seen an example of zig-zag riveting.

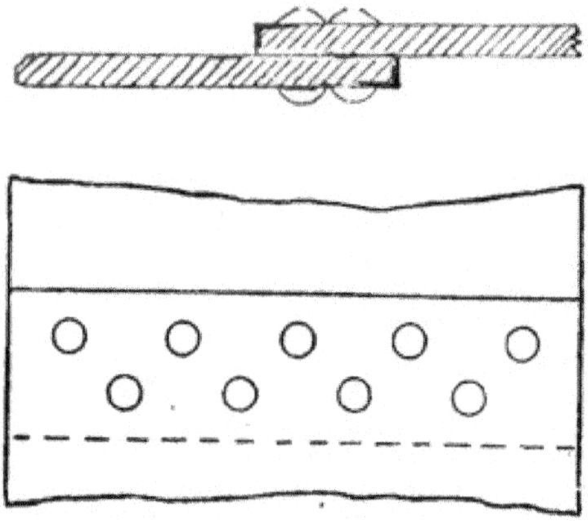

Fig. 41.

CAULKING—By this is meant the closing of the edges of the seams of boilers or plates. In preparing the seams for caulking, the edges are first planed true inside and outside; and after the plates have been riveted together, the edges are caulked or closed by a blunt chisel about ¼ inch thick at the edge which should be struck with a 3 or 4 lb. hammer; sometimes one man doing the work alone and sometimes one holding the chisel and another striking.

Fullering a boiler plate is done by a round-nosed tool, while *caulking* is executed by a sharper instrument.

The thinnest plate which should be used in a boiler is one-fourth of an inch, on account of the almost impossibility of caulking the seams of thinner plates.

It is a rule well known to all practical boiler makers that the thinner the metal (compatible with due strength) the longer the life of the boiler under its varying stresses and the better the caulking will stand.

STEEL RIVETS.

Hitherto there has been some prejudice against steel rivets, and while this may have some foundation when iron plates are used, it is certainly baseless when steel plates are concerned. The United States government has clearly demonstrated this. All the ships of the new navy have steel boilers, riveted with steel rivets, and an examination of the character of the material prescribed and the severity of the tests to which it is subjected show that these steel-riveted steel boilers are probably the best boilers ever constructed.

United States Government Requirements for Boiler Rivets.

They are subjected to the most severe hammer tests, such as flattening out cold to a thickness of one-half the diameter, and flattening out hot to a thickness of one-third the diameter. In neither case must they show cracks or flaws.

Kind of Material.—Steel for boiler rivets must be made by either the open-hearth or Clapp-Griffith process, and must not show more than .035 of one per centum of phosphorus nor more than .04 of one per centum of sulphur, and must be of the best quality in other respects.

Each ton of rivets from the same heat or blow shall constitute a lot. Four specimens for tensile tests shall be cut from the bars from which the lot of rivets is made.

Tensile Tests.—The rivets for use in the longitudinal seams of boiler shells shall have from 58,000 to 67,000 pounds tensile strength, with an elongation of not less than 26 per centum; and all others shall have a tensile strength of from 50,000 to 58,000 pounds, with an elongation of not less than 30 per centum in eight (8) inches.

Hammer Test.—From each lot twelve (12) rivets are to be taken at random and submitted to the following tests:

Four (4) rivets to be flattened out cold under the hammer to a thickness of one-half the diameter without showing cracks or flaws.

Four (4) rivets to be flattened out hot under the hammer to a thickness of one-third the diameter without showing cracks or flaws—the heat to be the working heat when driven.

STEEL RIVETS.

Four (4) rivets to be bent cold into the form of a hook with parallel sides without showing cracks or flaws.

Surface Inspection.—Rivets must be true to form, free from scale, fins, seams and all other unsightly or injurious defects.

In view of the fact that the government is using many hundred tons of these rivets, shown by the records of the tests to be vastly superior to any iron rivet made, in all the essentials of a good rivet, it would seem that it would benefit the boiler maker, the purchaser of the boiler and also the maker of the rivet by adopting a standard steel rivet to be used in all steel boilers.

BRACING OF STEAM BOILERS.

The material of a boiler being satisfactory and the plates being thoroughly and skillfully riveted there remains the important matter of strengthening the boiler against the enormous internal pressure not altogether provided for.

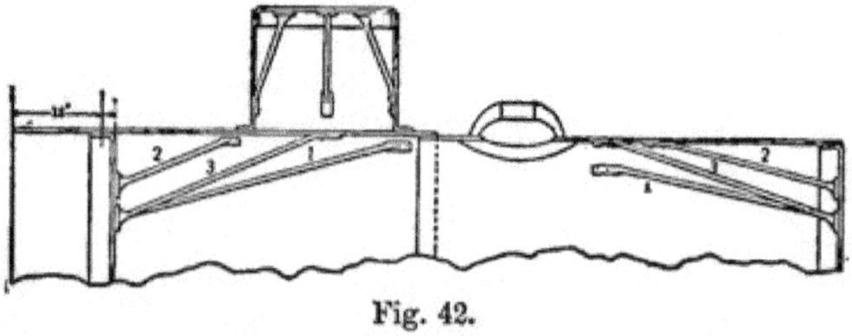

Fig. 42.

To illustrate the importance of attention to this point it may be remarked that a boiler eighteen feet in length by five feet in diameter, with 40 four-inch tubes, under a head of 80 pounds of steam, has a pressure of nearly 113 tons on each head, 1,625 tons on the shell and 4,333 tons on the tubes, making a total of 6,184 tons on the whole of the exposed surfaces.

Not only is this immense force to be withstood, but owing to the fact that the boiler grows weak with age—*a safety factor* of six has been adopted by inspectors, *i. e.*, the boiler must be made six times as strong as needed in every day working practice.

BRACING OF STEAM BOILERS.

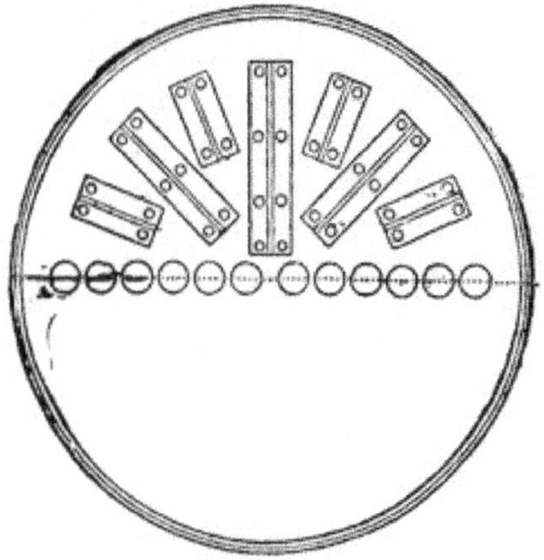

Fig. 43.

BRACES IN THE BOILER.—The proper bracing of flat surfaces exposed to pressure, is a matter of the greatest importance, as the power of resistance to bulging possessed by any considerable extent of such a surface, made as they must be in the majority of cases of thin plates, is so small that *practically the whole load has to be carried by the braces.* This being the case, it is evident that as much attention should be given to properly designing, proportioning, distributing and constructing the brace as to any other portion of the boiler.

All flat surfaces should be strongly supported with braces of the best refined iron, or mild steel, having a tensile strength of not less than 58,000 lbs. to the square inch. These braces must be provided with crow feet or heavy angle iron properly distributed throughout the boiler.

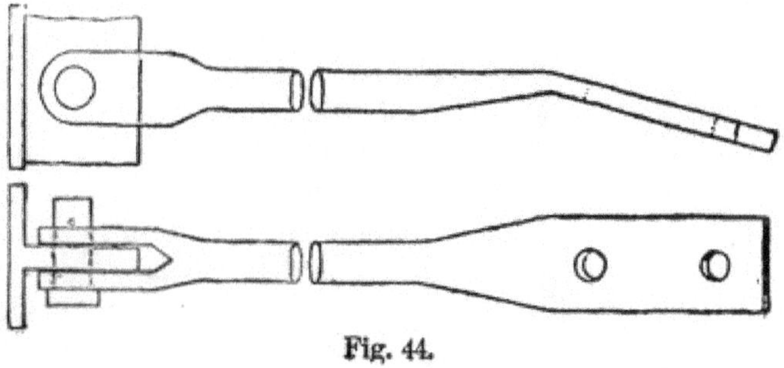

Fig. 44.

BRACING OF STEAM BOILERS.

Fig. 42 shows the method usually followed in staying small horizontal tubular boilers. The cut represents a 36 inch head and there are five braces in each head: two short ones and three long ones The braces should be attached to shell and head by two rivets at each end. The rivets should be of such size that *the combined area* of their shanks will be at least equal to the body of the brace, and their length should be sufficient to give a good large head on the outside to realize strength equal to the body of the brace.

In boilers with larger diameters, 5 to 8 feet, stay ends are made of angle or T iron; by this arrangement the stays can be placed further apart, the angle irons very effectively staying the plate between the stays, and thus affording more room in the body of the boiler. The size of the stays have to be increased in proportion to the greater load they have to sustain. See Fig. 43.

In a 66 inch boiler it is proper to have not less than 10 braces in each head, none under three feet in length, made of the best round iron one inch in diameter, with ends of braces made of iron 2½x½ inches with three pieces of T iron riveted to head above the tubes to which the braces are attached with suitable pins or turned bolts. See Fig. 44.

STAYING OF FLAT SURFACES.—When Boilers are formed principally of flat plates, like low-pressure marine boilers, or the fire-boxes of locomotive boilers, the form contributes nothing to the strength, which must, therefore, be provided for by staying the opposite furnaces together. Fig. 45 shows the arrangement of the stays in a locomotive fire-box. They are usually pitched about 4 inches from centre to centre, and are fastened into the opposite plates by screwing, as shown, the heads being riveted over. Each stay has to bear the pressure of steam on a square *aa*, and the sectional area of the stay must be so chosen that the tensile strength will be sufficient to bear the strain with the proper factor of safety.

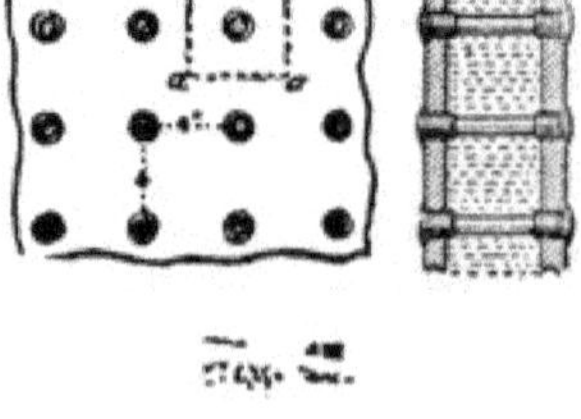

Fig. 45.

BRACING OF STEAM BOILERS.

If the spaces between the stays are too great, or the plate too thin, there is a danger of the structure yielding through the plate bulging outwards between the points of attachment of the stays, thus allowing the latter to draw through the screwed holes made in the plates.

In designing boilers with stayed surfaces, care should be taken that *the opposite plates connected by any system of stays should, as far as possible, be of equal area,* otherwise there is sure to be an unequal distribution of load in the stays, some receiving more than their proper share, and moreover, the least supported plate is exposed to the danger of buckling.

Rule for Finding Pressure or Strain on Bolts.

The absolute stress or strain on a flat surface of a steam boiler, which is carried by the stays, can be easily determined by a simple rule:

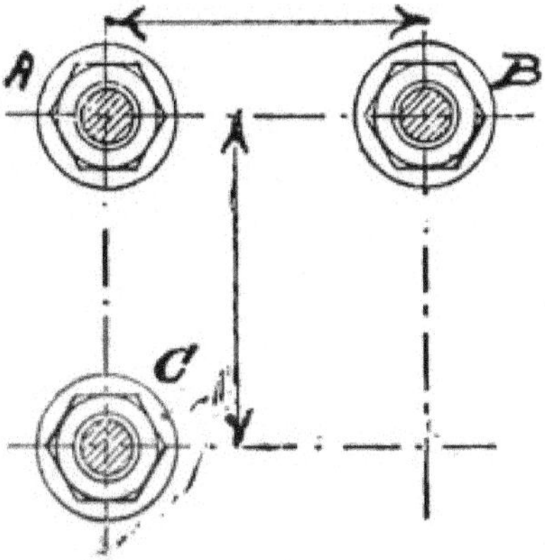

Fig. 46.

Choose 3 stays as A B C in Fig. 46, measure from A to B *in inches,* and from A to C. Multiply these two numbers together and the result is the number of square inches of surface depending upon one bolt for supporting strength.

Example.

Suppose the stays measure from center to center 5 inches each way with steam at 80 lbs., then

$$5 \times 5 = 25 \times 80 = 2{,}000 \text{ lbs. borne by 1 stay.}$$

Note.

The pressure on the surface does not include the space occupied by the area of the stay bolt, hence, to be absolutely correct that must be deducted.

GUSSET STAYS.

The flat ends of cylindrical boilers are, especially in marine boilers, stayed to the round portions of triangular plates of iron called gusset stays. These are simply pieces of plate iron secured to the boiler front or back, near the top or bottom, by means of two pieces of angle iron, then carried to the shell plating, and again secured by other pieces of angle bar. This arrangement is shown in Fig. 47.

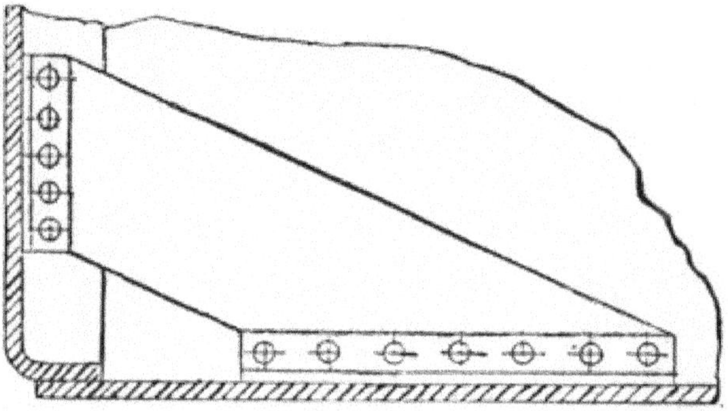

Fig. 47.

PALM STAYS.—These are shown in Fig. 48, and are often used in the same position as a gusset stay; that is, from the back or front end of the boiler to the shell plates; they are sometimes used to stay the curved tops of combustion chambers.

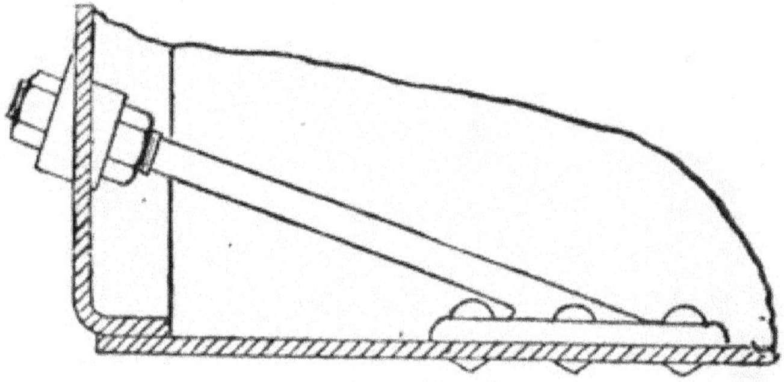

Fig. 48.

The two opposite ends are also stayed together by long bar stays, running the whole length of the boiler, it is dangerous, however, to trust too much to the latter class of stays; for, in consequence of the alternate expansion and contraction which

SCREWED STAYS.

takes place every time the boiler is heated and cooled, they have a tendency to work loose at the joints; and if the portion of the boiler in which they are situated should happen to be hotter than the outside shell, they have a tendency to droop, and are then perfectly useless.

RIVETED OR SCREW STAYS.

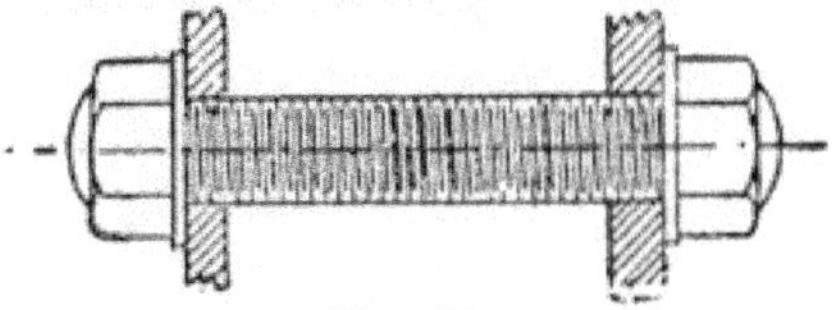

Fig. 49.

In addition to palm and gusset stays there are in use riveted or screwed stays, as shown in Fig. 49.

This would not answer in furnaces, owing to the burning off of the heads, hence driven stays are used there

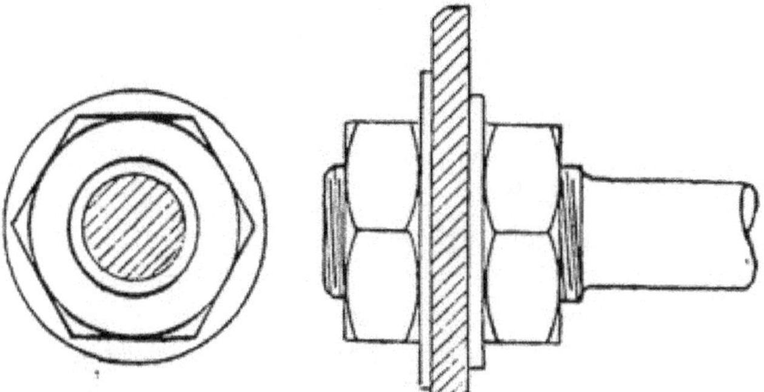

Fig. 50.

These screwed stays, shown in Fig. 50, are used (in marine and similar boilers) between the combustion chamber back and boiler back and also between the sides of the combustion chambers.

The general plan is to have a large nut and washer inside and outside the boiler with the outside washer considerable larger than the inside, so as to hold more efficiently the back and front ends together.

In marine boilers it is customary to place the stays 15 to 18 inches apart for ease of access to the parts of the boiler, and to make them of $2\frac{1}{4}$ to $2\frac{1}{2}$ inch iron of the best quality.

INSPECTOR'S RULES RELATING TO BRACES IN STEAM BOILERS, ALSO TO BE OBSERVED BY ENGINEERS.

Where flat surfaces exist, the inspector must satisfy himself that the spacing and distance apart of the bracing, and all other parts of the boiler, are so arranged that all will be of not less strength than the shell, and he must also after applying the hydrostatic test, thoroughly examine every part of the boiler.

No braces or stays employed in the construction of marine boilers shall be allowed a greater strain than six thousand pounds per square inch of section, and no screw stay bolt shall be allowed to be used in the construction of marine boilers in which salt water is used to generate steam, unless said stay bolt is protected by a socket. But such screw stay bolts, without sockets, may be used in staying the fire boxes and furnaces of such boiler, and not elsewhere, when fresh water is used for generating steam in said boiler. Water used from a surface condenser shall be deemed fresh water. And no brace or stay bolt used in a marine boiler will be allowed to be placed more than eight and one-half inches from centre to centre, except that flat surfaces, other than those on fire boxes, furnaces and back connections, may be reinforced by a washer or T iron of such size and thickness as would not leave such flat surface unsupported at a greater distance, in any case, than eight and one-half inches, and such flat surface shall not be of less strength than the shell of the boiler, and able to resist the same strain and pressure to the square inch, and no braces supporting such flat reinforced surfaces, will be allowed more than 16 inches apart.

In allowing the strain on a screw stay bolt, the diameter of the same shall be determined by the diameter at the bottom of the thread. Many State laws and City ordinances allow a strain of seven thousand five hundred pounds per square inch of section on good bracing without welds. The following table gives the safe load of round iron braces or stays.

DIAMETER OF BRACE.

Tensile strength per square inch of section allowed	½"	⅝"	¾"	⅞"	1"	1⅛"	1¼"	1½"	1¾"	2"
5000	981	1533	2208	3006	3927	4970	6136	8835	12026	15708
6000	1178	1840	2650	3607	4712	5964	7363	10602	14431	18849
7000	1374	2567	3092	4209	5497	6958	8590	12369	16837	21991
7500	1472	2730	3313	4509	5890	7455	9204	13253	18039	23562

SHOP NAMES FOR BOILER BRACES.—1. Gusset brace (fig. 47). 2. Crowfoot brace. 3. Jaw brace (fig. 44). 4. Head to head brace (fig. 50). These shop terms refer to braces used in the tubular form of boiler.

A STAY AND A BRACE in a steam boiler fulfil the same office, that of withstanding the pressure exerted outward of the expanded and elastic steam.

SOCKET BOLTS are frequently used instead of the screw stay between the inside and outside plates that form the center space. Socket bolts are driven hot the same as rivets.

The method of bracing with T bars is considered the best; the bars make the flat surface rigid and unyielding even before the brace is applied. The braces should be spaced about 8 inches apart on the T bar and 7 inches from the edge of the flange T the bar should be 4"×4"½" T iron and riveted to the head or flat surface with 1⅛" rivets spaced 4½ inches apart.

HOLLOW STAY BOLTS are used in the side of locomotive boilers at the top of the fire line, to aid the combustion; these are ordinarily 1¼" in diameter.

The flange of a boiler head ½" thick will amply support 6 inches from the edge of the flange.

A radius of 2 inches is ample for bend of flange on the head. The lower braces should be started 6 inches above the top row of tubes. Braces should be fitted so as to have a straight pull, *i. e.* parallel with the boiler shell. The heads of the boiler should be perfectly straight before the braces are fitted in place. Gusset brace plates should not be less than 30 inches long and 14 inches wide. Braces are best made of 1 inch O iron of highest efficacy with tensile strength of not less than 58,000 lbs. to the square inch.

POINTS RELATING TO BOILER BRACES.

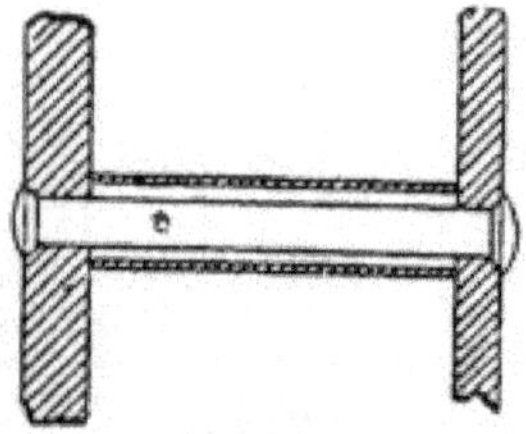

Fig. 51.

The riveted stay shown in Fig. 51, consists of a long rivet, passed through a thimble or distance piece of wrought iron pipe placed between plates, to be stayed together, and then riveted over in the usual manner.

An ingenious device is in use to show when a bolt has broken. A small hole is drilled into the head, extending a little way beyond the plate, and as experience shows that the fracture nearly always occurs *next to the outside plate*, that is the end taken for the bored out head: when the bolt is broken the rush of steam through the small hole shows the danger without causing serious disturbance.

Even where the best of iron is used for stay bolts they should never be exposed to more than $\frac{1}{10}$th or $\frac{1}{15}$th their breaking strength.

The stays should be well fitted, and each one carefully tightened, and, as far as possible each stay in a group *should have the same regular strain upon it*—if the "pull" all should come on one the whole are liable to give way.

Dimensions and Shape of Angle and T Iron.

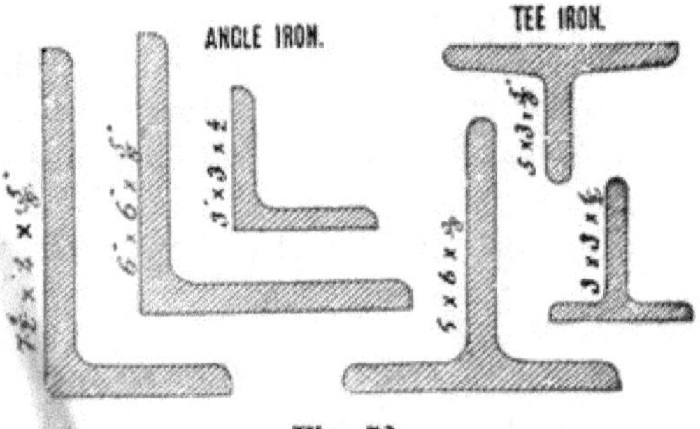

Fig. 52.

POINTS RELATING TO BOILER BRACES.

The condition of a boiler can be learned by tapping on the sheets, rivets, seams, etc., to ascertain whether there are any broken stays, laminated places, broken rivets, etc.

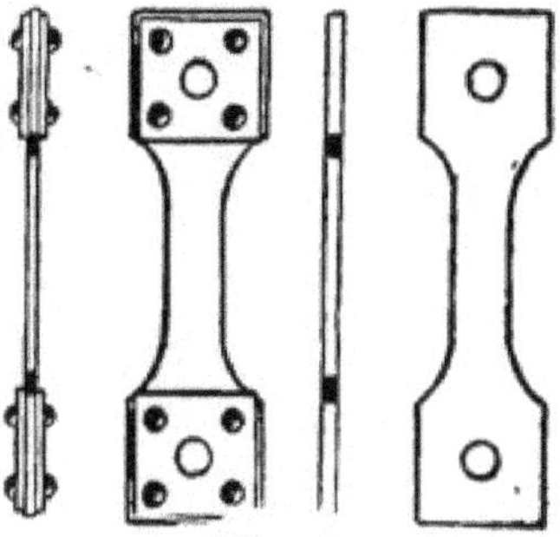

Fig. A.

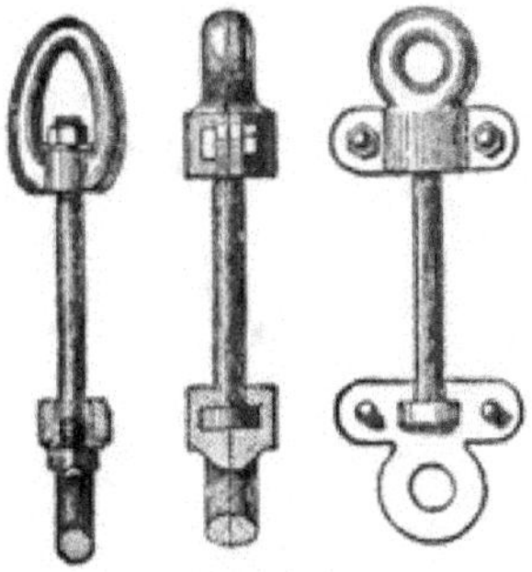

Fig. B.

Fig. A represents the method of preparing testing pieces of boiler plate, for the machines prepared specially to measure their elongation before breaking, and also the number of pounds they will bear stretching before giving way. Fig. B exhibits the same with reference to the brace and other O iron.

RULES AND TABLES

FOR DETERMINING AREAS AND CALCULATING THE CONTENTS OF STEAM AND WATER SPACES IN THE STEAM BOILER.*

To make these calculations, a circle should be drawn representing the circumference of the head of the boiler, and a line drawn across between points, corresponding with the ends of the upper row of tubes. Measure carefully that portion of the circle which is above these points which are represented by the figure in the diagrams C and D, and multiply it by one-quarter of the diameter of the circle. Then measure the length of the line 1, 2, multiply it by one-half of the dotted line drawn from the center of the circle to the base of the segment, and subtract this product from the result first obtained. The remainder will be the area of the segment.

*We are indebted to W. H. Wakeman, M. E., for this rule.

STEAM AND WATER SPACES.

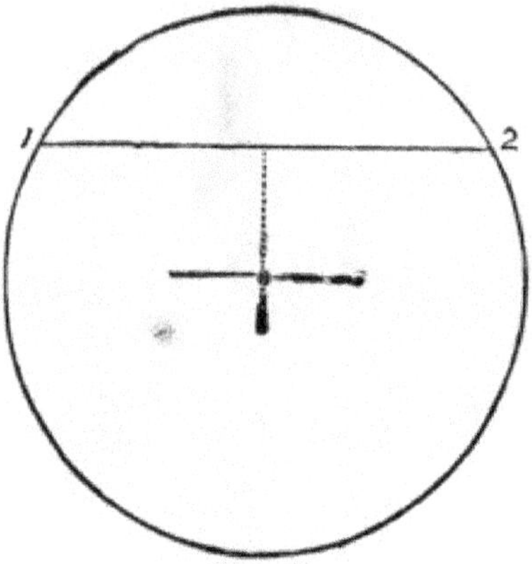

Fig. C.

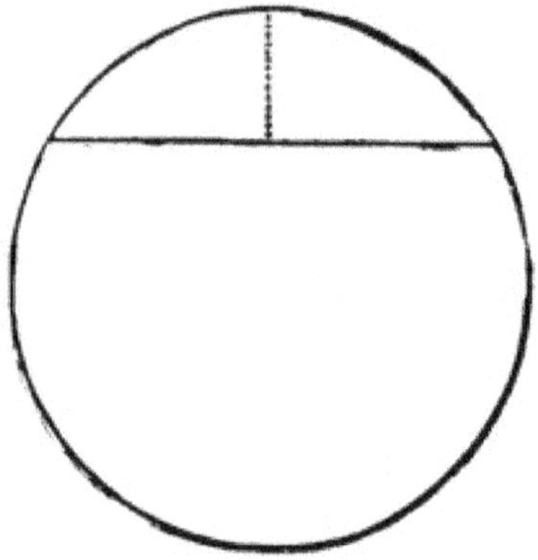

Fig. D.

Suppose the circle to be forty-eight inches in diameter and the segment fourteen inches high, the upper part of the circle between 1 and 2 will measure four feet six inches, or in exact figures, 54.6875 inches; one-quarter of the diameter of the circle is 12 inches and 54.6875×12=656.25. The straight line, from 1 to 2, we find to be, say, 43 75 inches in length, the line extending from the center of the circle to the segment base is 10 inches long, half of this is 5.0×43.75=218.75.

656—218.75=437.5,

the area in square inches of the segment.

To find the area of the larger portion of the circle, the length of the line from 1 to 2 must be carefully ascertained and multiplied by one-fourth the diameter of the circle. Half the length of the straight line from 1 to 2 must then be multiplied by one-half the dotted line, the product added to the figures already obtained, and the result will be the area of the larger portion of the circle. Special care must be bestowed on the measurement of the curved lines, as a mistake of a fraction of an inch will throw the calculations out.

For the following valuable tables we are indebted to successive numbers of that unique and instructive journal, the *Locomotive.*

TABLES FOR CALCULATING NUMBER OF STAYS.

The accompanying tables will greatly facilitate the calculation of the number of braces required in a boiler that is to run under any given pressure. They contain the results of long experience on the subject, and can be relied upon to give perfectly satisfactory results.

It has been shown by direct experiment that the tubes possess sufficient holding power to amply stay the part of the head to which they are attached, and we may safely consider that they will also possess sufficient staying power to take care of the head *for two* (2) *inches above their upper surfaces.*

The flanges of the heads being securely united to the shell, and being also curved or dished, it may likewise be safely assumed that *no braces need be provided for that part of the head which lies within three* (3) *inches of the shell.* The part to be braced, therefore, consists of a segment of a circle *whose circumference lies three inches within the circle of the shell, and whose base is two inches above the upper row of tubes.*

Thus in a 66-inch boiler, whose upper row of tubes is 26 inches below the top of the shell, the part of the head that requires bracing consists in a segment of a circle, the diameter of which is 60 inches, and the height of which is 21 inches; 21 inches being the measured height (26 in.), minus the 3 inches that lies between the shell and the segment to be braced, and minus the two inches that lies between this segment and the top of the tubes.

Table No. 1 gives the total area in square inches. No. 2, areas to be braced. No. 3, number of braces of one inch round iron required, allowing seven thousand five hundred pounds per square inch of section at one hundred pounds steam pressure.

Table 3 will be found of more practical use than Table 2, for it gives directly the number of braces required in any given boiler, instead of the area to be braced. It was calculated from Table 2. The iron used in braces will safely stand

TABLES FOR CALCULATING NUMBER OF STAYS.

a continuous pull of 7,500 pounds to the square inch, which is the figure used in computing the foregoing table. A round brace an inch in diameter has a sectional area of .7854 of an inch, and the strain that it will safely withstand is found by multiplying .7854 by 7,500, which gives 5,890 pounds as the safe working strain on a brace of one-inch round iron.

In a 60-inch boiler, whose upper tubes are 28 inches below the shell, the area to be braced is, according to table, [illegible],930 square inches. If the pressure at which it is to be run is 100 pounds to the square inch, the entire pressure on the area to be braced will be 93,000 pounds, and this is the strain that must be withstood by the braces. As one brace of inch-round iron will safely stand 5,890 pounds, the boiler will need as many braces as 5,890 is contained in 93,000, which is 15.8. That is, 16 braces will be required. The table is made out on the basis of 100 lbs. pressure to the square inch, because that is a very convenient number.

TABLE No. 1. TOTAL AREA ABOVE TUBES OR FLUES.

(SQUARE INCHES.)

Height from tubes to shell.	DIAMETER OF BOILER IN INCHES.							Height from tubes to shell.
	36	42	48	54	60	66	72	
15	389							15
16	419							16
17	458	526						17
18		566	620	667				18
19		608	667	720				19
20		650	714	770	824			20
21			756	824	882			21
22			808	878	937			22
23				930	996	1059		23
24				982	1056	1121		24
25				1037	1116	1184		25
26				1090	1209	1252	1324	26
27				1145	1234	1316	1394	27
28					1291	1381	1465	28
29					1352	1445	1536	29
30					1414	1511	1608	30
31						1576	1674	31
32						1641	1746	32
33							1818	33
34							1896	34

TABLES FOR CALCULATING NUMBER OF BRACES.

TABLE 2. AREAS TO BE BRACED (SQUARE INCHES.)

Height from tubes to shell.	DIAMETER OF BOILER IN INCHES.							Height from tubes to shell.
	36	42	48	54	60	66	72	
15	206							15
16	235							16
17	264	297						17
18		331	365	396				18
19		816	404	439				19
20		401	444	483	519			20
21			485	528	568			21
22			526	574	618			22
23				620	668	714		23
24				667	720	769		24
25				714	772	825		25
26				761	824	882	937	26
27				809	877	940	998	27
28					930	998	1061	28
29					983	1056	1124	29
30					1037	1115	1187	30
31						1174	1252	31
32						1234	1317	32
33							1382	33
34							1447	34

TABLE 3. NUMBER OF BRACES REQUIRED, AT 100 LBS. PRESSURE.

Height from tubes to shell.	DIAMETER OF BOILER IN INCHES.							Height from tubes to shell.
	36	42	48	54	60	66	72	
15	3.5							15
16	4.0							16
17	4.5	5.0						17
18		5.6	6.2	6.7				18
19		6.2	6.9	7.5				19
20		6.8	7.5	8.2	8.9			20
21			8.2	9.0	9.6			21
22			8.9	9.8	10.5			22
23				10.5	11.3	12.1		23
24				11.3	12.2	13.1		24
25				12.1	13.1	14.0		25
26				12.9	14.0	15.0	15.9	26
27				13.7	14.9	16.0	16.9	27
28					15.8	16.9	18.0	28
29					16.7	17.9	19.1	29
30					17.6	18.9	20.2	30
31						19.9	21.3	31
32						21.0	22.4	32
33							23.5	33
34							[illegible]	34

BOILER TUBES.

In Table 2 this calculation has been made for all sizes of boilers that are ordinarily met with. The area to be braced has been calculated as above in each case, the two-inch strip above the tubes, and the three-inch strip around the shell being taken into account. As an example of its use, let us suppose that upon measuring a boiler we find that its diameter is 54 inches, and that the distance from the upper tubes to the top of the shell is 25 inches. Then by looking in the table under 54″ and opposite 25″ we find 714, which is the number of square inches that requires staying on each head.

BOILER TUBES.

Table.

Dimensions of Lap Welded Boiler Tubes.

Size outside diameter.	Wire Guage.	Weight per foot.	Size outside diameter.	Wire Gauge.	Weight per foot.
1 inch.	15	0.708	3½ inches.	11	4.272
1¼ "	15	0.9	3¾ "	11	4.590
1½ "	14	1.250	4 "	10	5.320
1¾ "	13	1.665	4½ "	10	6.010
2 "	13	1.981	5 "	9	7.226
2¼ "	13	2.238	6 "	8	9.346
2½ "	12	2.755	7 "	8	12.435
2¾ "	12	3.045	8 "	8	15.109
3 "	12	3.333	9 "	7½	
3¼ "	11	3.958	10 "	6½	

The above is the regular manufacturers' list of sizes and weights.

Note.

Boiler tubes are listed and described from the *outside diameter.* This should be noted, as gas-pipe is described from the *inside* diameter. Thus a 1-inch gas-pipe is nearly 1¼ outside diameter while a 1-inch boiler tube is exactly one inch. Another difference between the two consists in the fact that the outside of boiler tubes is rolled smooth and even, gas-pipe is left comparatively rough and uneven.

BOILER TUBES.

When the boiler tubes are new and properly expanded there is a large reserve or surplus of holding power for that part of the tube sheet supported by them, this has been proved by experiment made by chief engineer W. H. Stock, U. S. N. as shown in the following

TABLE OF HOLDING POWER OF BOILER TUBES.

Outside diameter of end of tube where fracture took place.	Area of cross section of body of tube.	Thickness of tube plate.	Strain in pounds. Mean result.	Method of Fastening.
Inches.	Sq. ins.	Inches.	Pounds.	
$2\frac{5}{8}$	.981	$\frac{7}{16}$	22650	Expanded by Dudgeon tool, end riveted over.
$2\frac{5}{8}$	.981	$\frac{7}{16}$	22150	Expanded by Dudgeon tool, end partly riveted over.
$2\frac{5}{8}$	.981	$\frac{3}{8}$	25525	Expanded by Dudgeon tool, end riveted over.
$2\frac{5}{8}$	.981	$\frac{3}{8}$	29675	Expanded by Dudgeon tool, ferruled, not riveted over.
$2\frac{5}{8}$	.981	$\frac{3}{8}$	13050	Simply expanded by Dudgeon tool.

Mr. C. B. Richards, consulting engineer at Colt's Armory at Hartford, Conn., made some experiments as to the holding power of tubes in steam boilers, with the following results: The tubes were 3 inches in external diameter, and 0.109 of an inch thick, simply expanded into a sheet $\frac{3}{8}$ of an inch thick by a Dudgeon expander. The greatest stress without the tubes yielding in the plate was 4,500 pounds, and at 5,000 pounds was drawn from the sheet. These experiments were repeated with the ends of the tubes which projected through the sheet three-sixteenths of an inch, being flared so that the external diameter in the sheet was expanded to 3.1 inches. The greatest stress without yielding was 18,500 pounds; at 19,000 pounds yielding was observed; and at 19,500 pounds it was drawn from the sheet. The force was applied paralled to the axis of the tube, and the sheet surfaces were held at right angles to the tube axis.

BOILER TUBES.

NOTE.

When the tube sheet and tube ends near the the sheet become coated with scale or the tubes become over heated, the holding power of the tubes becomes largely reduced and caution must be used in having the tube ends re-expanded and accumulated scale removed.

NOTE. 2—In considering the stress or strain, upon the expanded or riveted over ends of a set of boiler tubes it may be remembered that the strain to be provided against is only that coming upon tube plate, exposed to pressure, *between the tube ends*—the space occupied by the tubes has no strain upon it.

The gauge to be employed by inspectors to determine the thickness of boiler plates, will be any standard American gauge furnished by the Treasury Department.

All samples intended to be tested on the Riehle, Fairbanks, Olsen, or other reliable testing machine, must be prepared in form according to the following diagram, viz: eight inches in length, two inches in width, cut out at their centers as indicated.

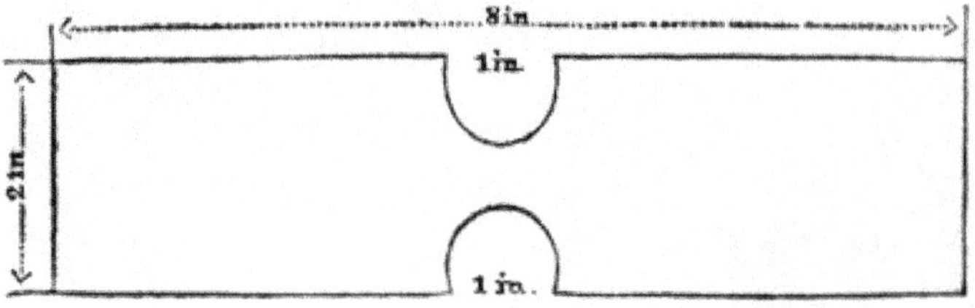

Fig. E.

PORTIONS OF THE MARINE BOILER WHICH BECOME THIN BY WEAR.

These are generally situated, 1st, at or a little above the line of fire bars in the furnace; 2d, the ash pits; 3d, combustion chamber backs; 4th, shell at water line; 5th, front and bottom of boiler.

The thinning can usually be detected by examination, sounding with a round nosed hammer, or drilling small holes in suspected parts not otherwise accessible for examination.

EXAMPLES OF CONSTRUCTION AND DRAWING

d	t	d	t
9/16	1/4	15/16	5/8
11/16	5/16	1 1/16	3/4
3/4	3/8	1 1/8	7/8
7/8	1/2	1 3/16	1

d = DIAM. OF RIVET.

t = THICKNESS OF PLATE.

The small table at the left is of use in this and the four succeeding pages; in all places in the drawings where "d" is used it indicates *the diameter of the rivet;* "t" means *the thickness of the plate;* "p" stands for *pitch.* The table also shows the proportion of rivet to the plate—thus, a ¼ inch plate requires a 9/16 rivet, etc.

It is recommended, in view of the increased disposition on the part of official examiners to test the applicant's knowledge of drawing, for any one interested, to redraw to *a full size* all the rivets, plates, and methods of joining the two contained on pages 113--116.

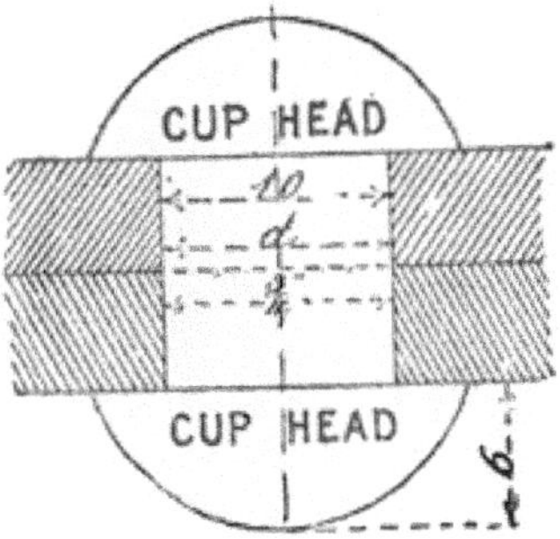

Fig. 53.

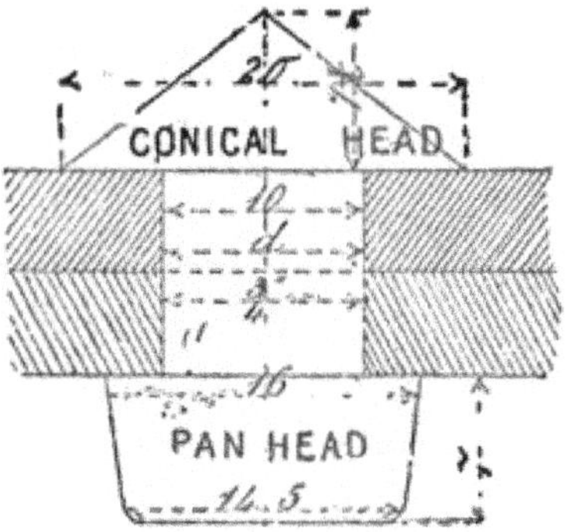

Fig. 54.

The figures 53 to 60 will be understood without much explanation.

In figures 53 and 54 *the cup head, the conical head and pan head rivets* are shown.

Figs. 55 and 56 exhibit the details (and drawings) of single and double riveting. Where the cut reads p=2½d, it means that that the distance from the center of one rivet to the center of the next shall be 2½ the diameter of the rivet, see example, page 115.

CONSTRUCTION AND DRAWING.

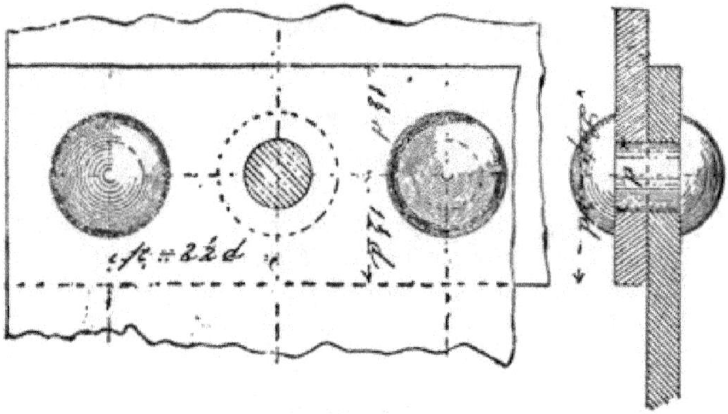

Fig. 55.

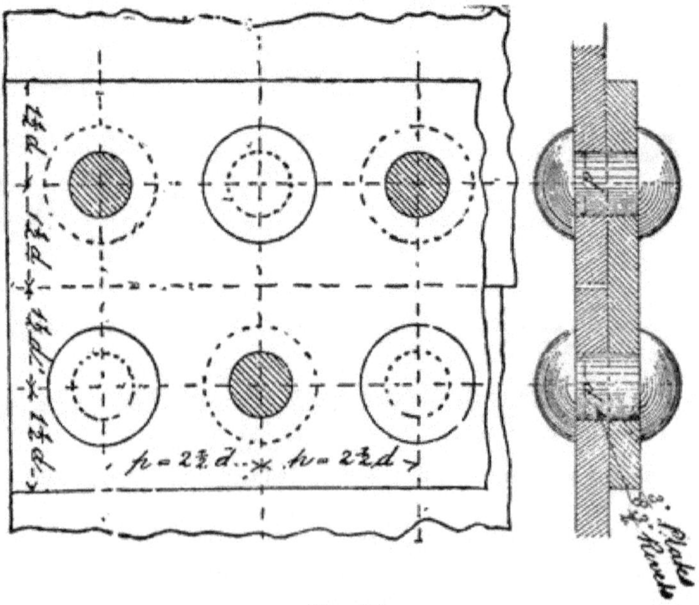

Fig. 56.

CONSTRUCTION AND DRAWING.

EXAMPLE.

If the size of the rivet used is $\frac{7}{8}$ths, then $\frac{7}{8}\times 2\frac{1}{2}=2\frac{3}{16}$ inches nearly, and this gives the proportionate strength of the plate and the rivet, see page 113.

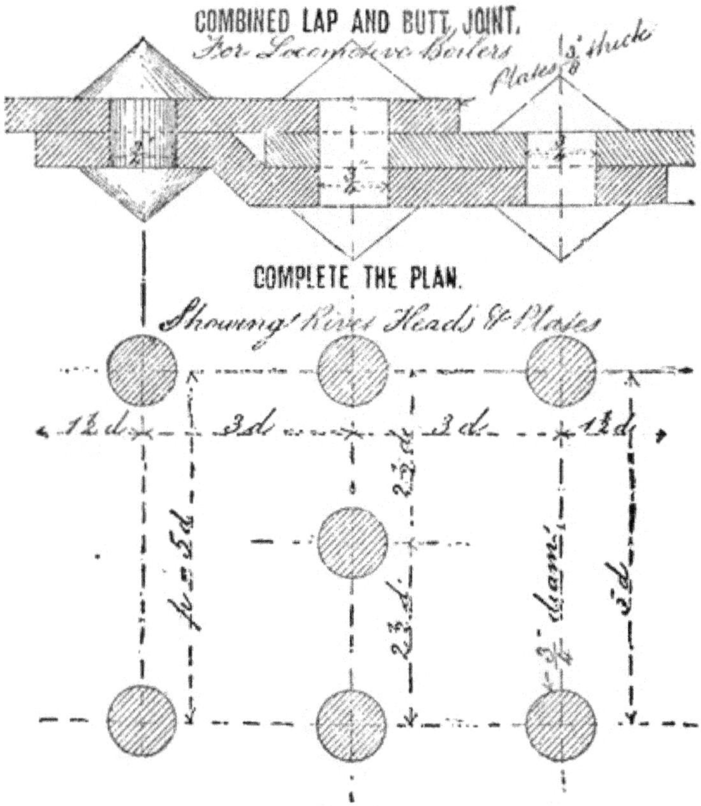

Fig. 57.

Figs. 57, 58, 59 and 60 show quite clearly the joints and rivet work done in locomotive and marine work. Fig. 60 shows method of riveting 3 plates, A, B, and C, together.

CONSTRUCTION AND DRAWING.

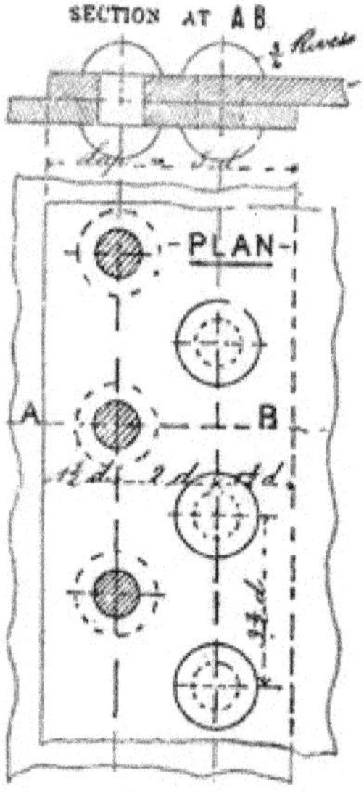

Fig. 58.

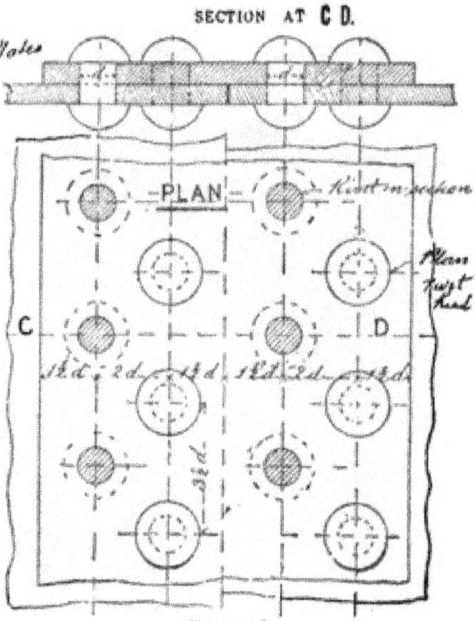

Fig. 59.

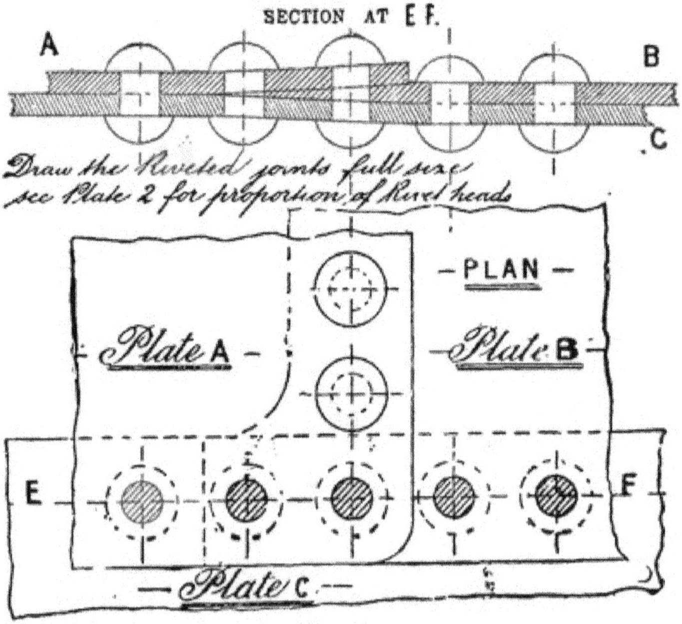

Fig. 60.

RULE FOR SAFE INTERNAL PRESSURE.

The safe internal pressure on cylindrical shells is found according to the following rule, which has been adopted by the United States Board of Supervising Inspectors, and any boiler shell not found in the tables can be determined by this rule.

RULE.—Multiply one-sixth of the lowest tensile strength found stamped on any plate in the cylindrical shell by the thickness—expressed in inches or parts of an inch—of the thinnest plate in the same cylindrical shell, and divide by the radius or half diameter—also expressed in inches—and the result will be the pressure allowable per square inch of surface for single-riveting, to which add twenty per centum for double riveting.

The hydrostatic pressure applied, under this table and rule, must be in the proportion of one hundred and fifty pounds to the square inch, to one hundred pounds to the square inch of the working pressure allowed.

EXAMPLE.

What pressure should be allowed to be carried on a boiler 60″ diameter, made of plates $\frac{3}{8}$″ thick, having a tensile strength of 60,000 pounds? Now then:

```
            6)60,000
              ------
              10,000
                   3
              ------
            8)30,000
              ------
Half diam.   30)3750(125 lbs.—if single riveted.
                30
                --
                 75
                 60
                ---
                150         125+25 lbs. (20 feet)=150 for
                150              double riveted.
```

TABLES SAFE INTERNAL PRESSURE.

Diameter of Boiler.	Thickness of Plates.	45,000 Tensile Strength. 1-6, 7,500. Pressure. Single Riveted.	45,000 Double Riveted.	50,000 Tensile Strength. 1-6, 8,333.3. Pressure. Single Riveted.	50,000 Double Riveted.	55,000 Tensile Strength. 1-6, 9,166.6. Pressure. Single Riveted.	55,000 Double Riveted.	60,000 Tensile Strength. 1-6, 10,000. Pressure. Single Riveted.	60,000 Double Riveted.	65,000 Tensile Strength. 1-6, 10,833.3. Pressure. Single Riveted.	65,000 Double Riveted.	70,000 Tensile Strength. 1-6, 11,666.6. Pressure. Single Riveted.	70,000 Double Riveted.	Thickness of Plates.
36 Inches.	.21	87.5	105.	97.21	116.65	106.94	128.3	116.66	139.99	126.38	151.65	136.11	163.33	.21
	.23	95.83	114.99	106.47	127.76	117.12	140.54	127.77	153.32	138.41	166.09	149 07	178.88	.23
	.25	104.16	124.99	115.74	138.88	127.31	152.77	138.88	166.65	150.46	180.55	162.03	194.43	.25
	.26	108.33	129.99	120.37	144.44	132.4	158.88	144.44	173.32	156.48	187.77	168.51	202.21	.26
	.29	120.83	144.99	134.25	161.11	147.68	177.21	161.11	193.33	174.53	209.43	187.90	225.48	.29
	.33	137.5	165.	152.77	183.32	168.05	201.66	183.33	219.99	198.61	238.33	213.88	256.65	.33
	.35	145.83	174.99	162.03	194.43	178.23	213.87	194.44	233.32	210.64	252.76	226 84	272.20	.35
	.375	156.25	187.5	173.61	208.33	190 97	229.16	208.33	249.99	225.69	271.82	243 05	291.66	.375
40 Inches.	.21	78.75	94.50	87.49	104.98	96.24	115.48	105.	126.	113.74	136.48	122.49	146.98	.21
	.23	86.25	103.5	95.83	114.99	105.41	126.49	115.	138.	124.58	149.49	134.16	160.99	.23
	.25	93.75	112.5	104.16	124.99	114.58	137.49	125.	150.	135.41	162.49	145.83	174.99	.25
	.26	97.5	117.	108.33	129.99	119.16	142.99	130.	156.	140.83	168.99	151.66	181.99	.26
	.29	108.75	130.5	120.83	144.99	132 91	159.49	145.	174.	157.08	188.49	169.16	202.99	.29
	.3125	117.18	140.61	130.2	156.24	143.22	171.86	156.25	187.45	169.27	203.12	182.29	218.74	.3125
	.33	123.75	148.5	137.49	164.98	151.24	181.48	165.	198.	178.74	214.48	192.49	230.98	.33
	.35	131.25	157.5	145.83	174.99	160.41	192.49	175.	210.	189.58	227 49	204.16	244.99	.35
	.375	140.62	168.74	156.24	187.48	171.87	206.24	187.5	225.	203.12	243.74	218 74	262.48	.375
42 Inches.	.21	75.	90.00	83.32	99.99	91.66	109.99	100.	120.	108.33	129.99	116.66	139.99	.21
	.23	82.14	98.56	91.23	109.51	100.39	120.46	109.52	131.42	118.65	142.38	127.77	153.22	.23
	.25	89.28	107.13	99.2	119.04	109.12	130.94	119.04	142 84	128.96	154.75	138.88	166.65	.25
	.26	92.85	111.42	103.17	123.8	113.49	136.18	123.8	148.56	134 12	160.94	144.44	173.32	.26
	.29	103.57	124.28	115.07	138.08	126.57	151.85	138.09	165.7	149.6	179.52	161.11	193.33	.29
	.3125	111.6	133.92	124.	148.8	136.4	163.68	148.74	178.56	161.2	193.44	173.61	208.23	.3125
	.33	117.85	141.42	130.94	157.12	144 04	172.84	157.14	188.56	170.23	204.27	183.33	219.99	.33
	.35	125.	150.	128.88	166.65	152.77	183.32	166.66	199.99	180.55	216.66	194.44	233.32	.35
	.375	133.92	160.7	148.8	178.56	163.68	196.40	178.57	214 28	193.45	232.14	208.33	249.99	.375

TABLES SAFE INTERNAL PRESSURE.

Diameter of Boiler	Thickness of Plate.	45,000 Tensile Strength. 1-6, 7,500. Pressure. Single Riveted.	45,000 Double Riveted.	50,000 Tensile Strength. 1-6, 8,333.3. Pressure. Single Riveted.	50,000 Double Riveted.	55,000 Tensile Strength. 1-6, 9,166.6. Pressure. Single Riveted.	55,000 Double Riveted.	60,000 Tensile Strength. 1-6, 10,000. Pressure. Single Riveted.	60,000 Double Riveted.	65,000 Tensile Strength. 1-6, 10,833.3. Pressure. Single Riveted.	65,000 Double Riveted.	70,000 Tensile Strength. 1-6, 11,666.6. Pressure. Single Riveted.	70,000 Double Riveted.	Thickness of Plates.
48 Inches.	.21	65.62	78.74	72.91	87.49	80.2	96.24	87.49	104.98	94.79	113.74	102.08	122.49	.21
	.23	71.87	86.24	79.85	95.82	87.84	105.4	95.83	114.99	103.81	124.57	111.8	133.16	.23
	.25	78.12	93.74	86.8	104.16	95.48	114.57	104.16	124.99	112.84	135.4	121.52	145.82	.25
	.26	81.25	97.50	90.27	108.32	99.3	119.16	108.33	129.99	117.36	140.83	126.38	151.65	.26
	.29	90.62	108.74	100.69	120.82	110.76	132.91	120.83	144.99	130.9	157.08	140.97	169.16	.29
	.3125	97.65	117.18	108.5	130.2	119.35	143.22	130.21	156.25	141.05	169.26	151.9	182.28	.3125
	.33	103.12	123.74	114.58	137.49	126.04	151.24	137.5	165.	148.95	178.74	160.41	192.49	.33
	.35	109.37	131.24	121.52	145.82	133.67	160.4	145.83	174.99	157.98	189.57	170.13	204.14	.35
	.375	117.18	140.61	130.2	156.24	143.22	171.86	156.25	187.50	169.27	203.12	182.29	218.74	.375
54 Inches.	.21	58.33	69.99	64.81	77.77	71.29	85.54	77.77	93.32	84.25	101.1	90.74	108.88	.21
	.23	63.88	76.65	70.98	85.17	78.08	93.69	85.18	102.21	92.38	110.73	99.38	119.25	.23
	.25	69.44	83.32	77.16	92.52	84.87	101.84	92.59	111.10	100.3	120.36	108.02	129.62	.25
	.26	72.22	86.66	80.24	96.28	88.27	105.92	96.29	115.54	104.31	125.17	112.44	134.8	.26
	.29	80.55	96.66	89.5	107.40	98.45	118.14	107.41	128.88	116.35	139.62	125.3	150.36	.29
	.3125	86.8	104.16	96.44	115.72	106.09	127.30	115.55	138.66	125.38	150.45	135.03	162.03	.3125
	.33	91.66	109.99	101.84	122.22	112.03	134.43	122.22	146.66	132.4	158.88	142.59	171.10	.33
	.35	97.22	116.66	108.03	129.62	118.82	142.58	129.69	155.54	140.43	168.51	151.23	181.47	.35
	.375	104.16	124.99	115.74	138.88	127.31	152.77	138.88	166.65	150.46	180.55	162.03	194.43	.375
60 Inches.	.21	52.5	63.	58.33	69.99	64.16	76.99	69.99	84.	75.83	90.99	81.66	97.99	.21
	.23	57.5	69.	63.88	76.65	70.27	84.32	76.66	91.99	83.05	99.66	89.44	107.32	.23
	.25	62.5	75.	69.44	83.32	76.38	91.65	83.33	99.99	90.27	108.32	97.22	116.66	.25
	.26	65.	78.	72.22	86.66	79.44	95.32	86.66	103.99	93.88	112.65	101.11	121.33	.26
	.29	72.5	87.	80.55	96.66	88.61	106.33	96.66	115.99	104.72	125.66	112.77	135.32	.29
	.3125	78.12	93.74	86.8	104.16	95.48	114.57	104.18	124.99	112.95	135.54	121.52	145.82	.3125
	.33	82.5	99.	91.66	109.99	100 83	120.99	109.99	132.	119.16	142.99	128.33	153.99	.33
	.35	87.5	105.	97.22	116.66	106.94	128.32	116.66	139.99	126.38	151.65	136.11	163.33	.35
	.375	93.75	112.5	104.16	124.99	114.58	137.49	125.	150.	135.41	162.49	145.83	174.99	.375

TABLES SAFE INTERNAL PRESSURE.

Diameter of Boiler.	Thickness of Plates.	45,000 Tensile Strength. 1-6, 7,500. Pressure. Single Riveted.	45,000 Double Riveted.	50,000 Tensile Strength. 1-6, 8,333.3. Pressure. Single Riveted.	50,000 Double Riveted.	55,000 Tensile Strength. 1-6, 9,166.6. Pressure. Single Riveted.	55,000 Double Riveted.	60,000 Tensile Strength. 1-6, 10,000. Pressure. Single Riveted.	60,000 Double Riveted.	65,000 Tensile Strength. 1-6, 10,833.3. Pressure. Single Riveted.	65,000 Double Riveted.	70,000 Tensile Strength. 1-6, 11,666.6. Pressure. Single Riveted.	70,000 Double Riveted.	Thickness of Plates.
66 Inches.	.1875	42.61	51.13	47.34	56.8	52.07	62.49	56.81	68.17	61.55	73.86	66.28	79.53	.1875
	.21	47.72	57.26	53.	63.63	58.33	69.99	63.63	76.35	68.93	82.71	74.24	89.08	.21
	.23	52.27	62.72	58.	69.69	63.88	76.65	69.69	83.63	75.5	90.6	81.31	97.57	.23
	.25	56.81	68.17	63.13	75.75	69.44	83.32	75.75	90.90	82.07	98.48	88.37	106.04	.25
	.26	59.09	70.9	65.65	78.78	72.22	86.66	78.78	94.53	85.35	102.42	91.91	110.29	.26
	.29	65.90	79.08	73.23	87.87	80.55	96.66	87.87	105.44	95.2	114.24	102.52	123.02	.29
	.3125	71.	85.2	78.91	94.69	86.89	104.16	94.69	113.62	102.58	123.09	110.47	132.56	.3125
	.33	75.	90.	83.33	99.99	91.66	109.99	99.99	120.	108.33	129.99	116.66	139.99	.33
	.35	79.56	95.47	88.38	106.05	97.22	116.66	106.	127.27	114.89	137.86	123.73	148.47	.35
	.375	85.22	102.26	94.69	113.62	104.16	124.99	113.62	136.34	123.1	147.72	132.57	159.08	.375
72 Inches.	.1875	39.06	46.87	43.4	52.08	47.74	57.28	52.08	62.49	56.42	67.70	60.76	72.91	.1875
	.21	43.75	52.5	48.6	58.33	53.47	64.16	58.33	69.99	63.19	75.82	68.05	81.66	.21
	.23	47.91	57.49	53.24	63.88	58.56	70.27	63.88	76.65	69.21	83.05	74.53	89.48	.23
	.25	52.08	62.49	57.87	69.44	63.65	76.38	69.44	83.32	75.22	90.26	81.01	97.21	.25
	.26	54.16	64.99	60.18	72.22	66.2	79.44	72.22	86.66	78.24	93.88	84.25	101.10	.26
	.29	60.41	72.49	67.12	80.55	73.84	88.60	80.55	96.66	87.26	104.71	93.98	112.77	.29
	.3125	65.10	78.12	72.33	86.8	79.57	95.48	86.8	104.16	94.03	112.83	101.27	121.52	.3125
	.33	68.75	82.5	76.38	91.62	84.02	100.82	91.66	109.99	99.3	119.16	106.94	128.32	.33
	.35	72.91	87.49	81.01	97.21	89.11	106.93	97.22	116.66	105.32	126.38	113.42	136.1	.35
	.375	78.12	93.74	86.8	104.16	95.48	114.57	104.16	124.99	112.84	135.43	121.52	145.82	.375

DEFINITION OF TERMS.

In the accompanying sections, some of the properties of iron and steel, as employed in the construction of boilers, are given. It is, therefore, desirable that the meanings applied to the various terms used should be clearly understood. The definitions necessary are, then, briefly as follows :—

Tensile Strength is equivalent to the amount of force which, steadily and slowly applied in a line with the axis of the test piece, just overcomes the cohesion of the particles, and pulls it into separate parts.

Contraction of Area is the amount by which the area, at the point where the specimen has broken, is reduced below what it was before any strain or pulling force was applied.

Elongation is the amount to which the specimen stretches, between two fixed points, due to a steady and slowly applied force, which pulls and separates it into parts. Elongation is made up of two parts; one due to the general stretch, more or less, over the length; the other, due to contraction of area at about the point of fracture.

Shearing strength is equivalent to the force which, if steadily and slowly applied at right angles, or nearly so, to the line of axis of the rivet, causes it to separate into parts, which slide over each other, the planes of the surface at the point of separation being at right angles, or nearly so, to the axis of the rivet.

Elastic limit is the point where the addition to the permanent set produced by each equal increment of load or force, steadily and slowly applied, ceases to be fairiy uniform, and is suddenly, after the point is reached, increased in amount. It is expressed as a percentage of the tensile strength.

Tough.—The material is said to be "tough" when it can be bent first in one direction, then in the other, without fracturing. The greater the angles it bends through (coupled with the number of time it bends), the tougher it is.

Ductile.—The material is "ductile" when it can be extended by a pulling or tensile force and remain extended after the force is removed. The greater the permanent extension, the more ductile the material.

DEFINITION OF TERMS.

Elasticity is that quality in a material by which, after being stretched or compressed by force, it apparently regains its original dimensions when the force is removed.

Fatigued is a term applied to the material when it has lost in some degree its power of resistance to fracture, due to the repeated application of forces, more particularly when the forces or strains have varied considerable in amount.

Malleable is a term applied to the material when it can be extended by hammering, rolling, or otherwise, without fracturing, and remains extended. The more it can be extended without being fractured, the more malleable it is.

Weldable is a term applied to the material if it can be united, when hot, by hammering or pressing together the heated parts. The nearer the properties of the material, after being welded, are to what they were before being heated and welded, the more weldable it is.

Cold-short is a name given to the material when it cannot be worked under the hammer or by rolling, or be bent when cold without cracking at the edges. Such a material may be worked or bent when at a great heat, but not at any temperature which is lower than about that assigned to dull red.

Hot-short is when the material cannot be easily worked under the hammer, or by rolling at a red-heat at any temperature which is higher than about that assigned to a red-heat, without fracturing or cracking. Such a material may be worked or bent at a less heat.

Homogeneous describes a material which is all of the same structure and nature.

A homogeneous material is the best for boilers, and it should be of suitable tensile strength with contraction of area and elongation best suited for the purpose, having an elastic limit that will insure the structure being reliable; it should be tough and ductile, and its elasticity fairly good, and be capable of enduring strains without becoming too quickly or easily fatigued. The material should be malleable and in some cases weldable; that which is of a decidedly cold-short or hot-short nature should be avoided.

BOILER REPAIRS.

This cut represents a form of clamp used in holding the plates against each other when being riveted.

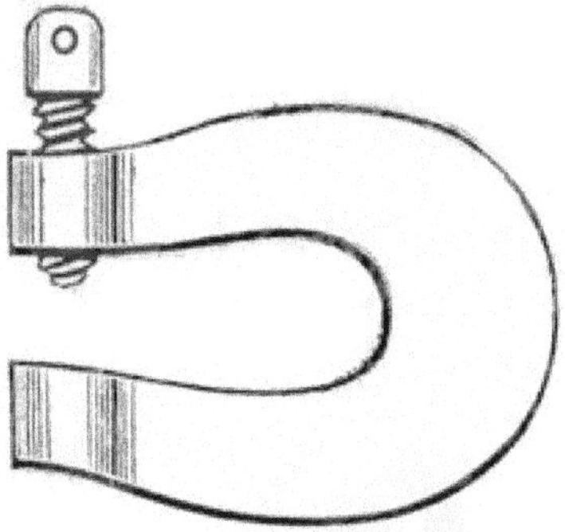

Fig. 66.

Fig. 67.

Fig. 67 represents a peculiar form of bolt for screwing a patch to a boiler. It is threaded into the boiler plate, the champer rests against the patch and the square is for the application of the wrench. After the bolt is well in place, the head can be cut off with a cold chisel.

REPAIRING CRACKS.

Cracks in the crown-sheet or side of a fire-box boiler, or top head of the upright boiler can be temporarily repaired by a row of holes drilled and tapped touching one another, with $\frac{3}{8}$ or $\frac{1}{2}$ inch copper plugs or bolts, screwed into the plates and afterwards all hammered together.

For a permanent job, cut out the defect and rivet on a patch. This had better be put on the inside, so as to avoid a "pocket" for holding the dirt. In putting on all patches, the defective part must be entirely removed to the solid iron, especially when exposed to the fire.

NOTE.—When fire comes to two surfaces of any considerable extent, the plate next to the fire becomes red-hot and weakens, hence the inside plate, in repairs, must be removed.

The application of steel patches to iron boilers is injudicious. Steel and iron differ structurally and in every other particular, and their expansion and contraction under the influence of changing temperatures, is such that trouble is sure to result from their combination.

DEFECTS AND NECESSARY REPAIRS.

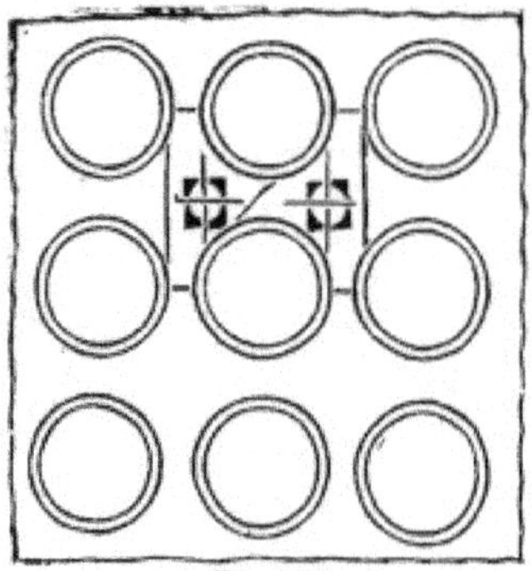

Fig. 68.

Fig. 68 represents a patch called a "spectacle piece." This is used to repair a crack situated between the tube ends. These are usually caused (if the metal is not of bad quality) by allowing incrustation to collect on the plate inside the boiler, or by opening the furnace and smoke doors, thus allowing a current of cold air to contract the metal of the plates round the heated and expanded tubes.

The "spectacle piece" is bored out to encircle the tubes adjacent to the crack, or in other words, to be a duplicate of a portion of the tube plate cracked. These plates are then pinned on to the tube covering the crack.

Steam generators, as they are exposed to more or less of trying service in steam production develop almost an unending number and variety of defects.

When a boiler is new and first set up it is supposed to be clean, inside and out, but even one day's service changes its condition; sediment has collected within and soot and ashes without.

Unlike animals and plants they have no recuperative powers of their own—whenever they become weakened at any point the natural course of the defect is to become continually worse.

In nothing can an engineer better show his true fitness than in the treatment of the beginnings of defects as they show themselves by well known signs of distress, such as leaks of water about the tube ends, and in the boiler below the water line, or escaping steam above it. In more serious cases, the professional services of a skillful and honest boiler maker is the best for the occasion.

DEFECTS AND NECESSARY REPAIRS.

In a recent report given in by the Inspectors the following list of defects in boilers coming under their observation was reported. The items indicate the nature of the natural decay to which steam boilers in active use are exposed. The added column under the heading of "dangerous" carries its own lesson, urging the importance of vigilance and skill on the part of the engineer in charge.

Nature of Defects.	Whole Number.	Dangerous.
Cases of deposit of sediment	419	36
Cases of incrustation and scale	596	44
Cases of internal grooving	25	16
Cases of internal corrosion	139	21
Cases of external corrosion	347	114
Broken and loose braces and stays	83	50
Settings defective	129	14
Furnaces out of shape	171	14
Fractured plates	181	84
Burned plates	93	31
Blistered plates	232	22
Cases of defective riveting	306	34
Defective heads	36	20
Serious leakage around tube ends	549	57
Serious leakage at seams	214	53
Defective water gauges	128	14
Defective blow-offs	45	9
Cases of deficiency of water	9	4
Safety-valves overloaded	22	7
Safety-valves defective in construction	41	16
Pressure-gauges defective	211	29
Boiler without pressure-gauges	3	0

This list covers nearly, if not all, *the points of danger* against which the vigilance of both engineer and fireman should be continually on guard; and is worth constant study until thoroughly memorized.

NOTE.

Probably one-quarter, if not one-third, of all boiler-work is done in the way of repairs, hence the advice of men who have had long experience in the trade is the one safe thing to follow for the avoidance of danger and greater losses, and for the best results the united opinion of 1, the engineer, experienced in his own boiler and 2, the boiler-maker with his wider observation and 3, the owner of the steam plant, all of whom are most interested.

DEFECTS AND NECESSARY REPAIRS.

Corrosion is a trouble from which few if any boilers escape. The principal causes of external corrosion arise from undue exposure to the weather, improper setting, or possibly damp brick work, leakage consequent upon faulty construction, or negligence on the part of those having them in charge.

Internal corrosion may be divided into ordinary corroding, or rusting and pitting. Ordinary corrosion is sometimes uniform through a large portion of the boiler, but is often found in isolated patches which have been difficult to account for. Pitting is still more capricious in the location of its attack ; it may be described as a series of holes often running into each other in lines and patches, eaten into the surface of the iron to a depth sometimes of one-quarter of an inch. Pitting is the more dangerous form of corrosion, and the dangers are increased when its existence is hidden beneath a coating of scale. There is another form of decay in boilers known as grooving ; it may be described as surface cracking of iron, caused by its expansion and contraction, under the influence of differing temperatures. It is attributable generally to the too great rigidity of the parts of the boiler affected, and it may be looked upon as resulting from faulty construction.

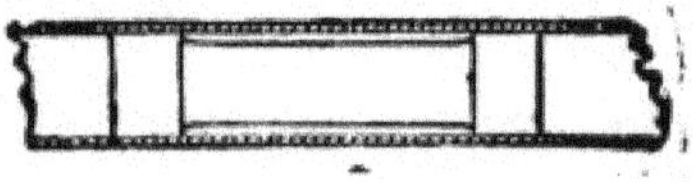

Fig. 69.

In plugging a leaky tube with a pine plug, make a small hole, of $\frac{3}{16}$ of an inch diameter, or less, running through it from end to end. These plugs should never have a taper of more than $\frac{1}{8}$ of an inch to the foot. It is well to have a few plugs always on hand. Fig. 69 exhibits the best shape for the wooden plug.

QUESTIONS

BY THE PROPRIETOR TO THE ENGINEER IN CHARGE, RELATING TO CONDITION OF THE BOILER.

How long since you were inside your boiler?

Were any of the braces slack?

Were any of the pins out of the braces?

Did all the braces ring alike ?

Did not some of them sound like a fiddle-string ?

Did you notice any scale on flues or crown sheet ?

If you did, when do you intend to remove it ?

Have you noticed any evidence of bulging in the fire-box plates ?

Do you know of any leaky socket bolts ?

Are any of the flange joints leaking ?

Will your safety valve blow off itself, or does it stick a little sometimes ?

Are there any globe valves between the safety valve and the boiler ? They should be taken out at once, if there are.

Are there any defective plates anywhere about your boiler ?

Is the boiler so set that you can inspect every part of it when necessary ?

If not, how can you tell in what condition the plates are ?

Are not some of the lower courses of tubes or flues in your boiler choked with soot or ashes ?

Do you absolutely know, of your own knowledge, that your boiler is in safe and economical working order, or do you merely suppose it is ?

QUESTIONS

ASKED OF A CANDIDATE FOR A MARINE LICENSE RELATING TO DEFECTS IN BOILER WITH ANSWERS.

If you find a thin plate, what would you do ?

Put a patch on.

QUESTIONS AND ANSWERS RELATING TO THE MARINE BOILER.

Would you put it on inside or outside?

Inside.

Why so?

Because the action that has weakened the plate will then act on the patch, and when this is worn it can be replaced; but the plate remains as we found it.

If the patch were put on the outside, the action would still be on the plate, which would in time be worn through, then the pressure of the steam would force the water between the plate and the patch, and so corrode it; and during a jerk or extra pressure, the patch might be blown off.

It is on the same principle that mud-hole doors are on the inside.

If you found several thin places, what would you do?

Patch each, and reduce the pressure.

If you found a blistered plate?

Put a patch on the fire side.

If you found a plate at the bottom buckled?

Put a stay through the centre of the buckle.

If you found several?

Stay each, and reduce the pressure.

The crown of the furnace down?

Put a stay through the middle, and a dog across the top.

If a length of the crown were down, put a series of stays and dogs.

A cracked plate?

Drill a hole at each end of the crack; caulk the crack, or put a patch over it.

If the water in the boiler is suffered to get too low, what may be the consequence?

Burn the top of the combustion chamber and the tubes; perhaps cause an explosion.

If suffered to get too high?

Cause priming; perhaps cause the breaking of the cylinder covers.

THE INSPECTION OF STEAM BOILERS.

Let it be clearly understood that if there were no steam generators using steam under pressure *there would be no boiler inspection, and no licensing of engineers;* it requires no license to be a machinist or a machine tender, no more would a license be essential to run a steam engine, except it were connected with the boiler. *The danger to the public arising from their use requires that the care and management of high pressure steam boilers shall be in hands of careful, experienced and naturally ingenious men,* hence it is on the affairs of the Boiler Room that the first tests are made, as to the worthiness of an aspirant for an engineer's license, hence too, the success of many firemen in obtaining the preference over engine-builders or school graduates, in the line of promotion as steam engineers.

The inspection laws of the various states and cities are framed after substantially the same leading ideas, and in presenting one the others may be assumed to be nearly the same.

The special province of the Steam Boiler Inspection and Engineers' Bureau in the police department in New York City is to inspect and test all the steam boilers in the city, at certain stated periods, and to examine every applicant for the position of engineer as to his ability and qualifications for running an engine and boiler with safety.

According to the laws of the State, every owner, agent or lessee, of a steam boiler or boilers, in the city of New York, shall annually report to the board of police, the location of said boiler or boilers, and, thereupon, the officers in command of the sanitary company shall detail a practical engineer, who shall proceed to inspect such steam boiler or boilers, and all apparatus and appliances connected therewith."

When a notice is received from any owner or agent that he has one or more boilers for inspection, a printed blank is returned to him stating that on the day named therein the boilers

INSPECTION OF STEAM BOILERS.

will be tested, and he is asked to make full preparation for the inspection by complying with the following rules :

Be ready to test at the above named time.
Have boiler filled with water to safety valve.
Have 1¼ inch connection.
Have steam gauge.
Steam allowed two-thirds amount of hydrostatic pressure.

More particularly stated, the following have been adopted by one or more Inspection Companies.

How to Prepare for Steam Boiler Inspection.

1. Haul fires and all ashes from furnaces and ash pits.

2. If time will permit, allow boiler and settings to cool gradually until there is no steam pressure, then allow water to run out of boilers. It is best that steam pressure should not exceed ten pounds if used to blow water out.

3. Inside of boiler should be washed and dried through manholes and handholes by hose service and wiping.

4. Keep safety valves and gauge cocks open.

5. Take off manhole and handhole plates as soon as possible after steam is out of boiler, that boiler may cool inside sufficiently for examination ; also *keep all doors shut* about boilers and settings, *except the furnace and ash pit doors.* Keep *dampers* open in *pipes* and *chimneys.*

6. Have all ashes removed from under boilers, and fire surfaces of shell and heads swept clean.

7. Have spare packing ready for use on manhole and handhole plates, if the old packing is made useless in taking off or is burned. The boiler attendant is to take off and replace these plates.

8. Keep all windows and doors to boiler room open, after fires are hauled, so that boilers and settings may cool as quickly as possible

INSPECTION OF STEAM BOILERS.

9. Particular attention is called to Rule 5, respecting doors—which should be open and which closed—also arrangement of damper. The importance of cooling the inside of the boiler by removal of manhole and handhole plates at the same time the outside is cooling, is in equalizing the process of contraction.

ISSUING CERTIFICATES.

These conditions having been complied with, the boiler is thoroughly tested, and if it is deemed capable of doing the work required of it, a number by which it shall hereafter be known and designated is placed upon it in accordance with the city ordinance: Failure to comply with this provision is punishable by a fine of $25. A certificate of inspection is then given to the owner, for which a fee of $2 is paid.

This certificate sets forth that on the day named the boiler therein described was subject to a hydrostatic pressure of a certain number of pounds to the square inch. The certificate tells where the boiler was built, its style or character and "now appears to be in good condition and safe to sustain a working pressure of —— to the square inch. The safety valve has been set to said pressure." A duplicate of this certificate is posted in full view in the boiler-room. In case the boiler does not stand the test to which it is subject, it must be immediately repaired and put in good working order before a certificate will be issued.

THE HYDRAULIC TEST.

The hydraulic test is a very convenient method of testing *the tightness of the work in a new boiler*, in conjunction with inspection to a greater or lesser degree, in the passing of new work. As a detector of leakages it has no rival, and its application enables faulty caulking to be made good before the boiler has left the works, and before a leak has time to enter on its insidious career of corrosion. The extent to which it enables the soundness and quality of the work to be ascertained is another matter, and depends on several conditions. It will be evident that if the test be applied with this object to a new boiler, the pressure should range to some point in excess of the

INSPECTION OF STEAM BOILERS.

working load if such a test is to be of any practical value. What the excess should be so as to remain within safe limits cannot be stated without regard being paid to the factor of safety adopted in the structure.

In addition to the advantage which the hydraulic test affords as a means of proving the tightness of the riveted seams and work generally, it is also of frequent assistance in determining the sufficiency of the staying of flat surfaces, especially when of indeterminate shape, or when the stresses thrown upon them by the peculiar construction of the boiler are of uncertain magnitude. For the hydraulic test, however, to be of any real value in the special cases to which we refer, it is essential that it should be conducted by an expert, and the application of the pressure accompanied by careful gaugings, so as to enable the amount of bulging and permanent set to be ascertained. Without such readings the application of the test in such cases is worthless, and may be delusive. Indeed, the careful gauging of a boiler as a record of its behavior should be a condition of every test, and is a duty requiring for its adequate performance a skilled inspector.

The duty of inspecting a new boiler or witnessing the hydraulic test properly belongs to one of the regular inspecting companies, who have men in their employ specially trained for the performance of such work. The advantage accruing from such a course is well worth the fee charged for the service, and secures a searching inspection of the workmanship, which frequently brings to light defects and oversights that a mere pumping-up of the boiler would never reveal. Such a proceeding in fact, can only prove that the boiler is water-tight, and a boiler may be tight under test although the workmanship is of the poorest character. Besides, it is well to bear in mind that the tightness of a boiler under test is no guarantee of its tightness after it is got to work. In a word, as far as new boilers are concerned, the application of hydraulic pressure unaccompanied by careful inspection and guagings may be almost worthless, while with these additions it may be extremely valuable, especially in the case of boilers of peculiar shape, and is a precaution that should not be neglected.

ENGINEERS' EXAMINATIONS.

Keeping in mind the fact that *if there were no steam-boilers there would be no examinations* and no public necessity for licenses, these "points" are added.

Examinations are trying periods with all engineers, as the best are liable to fail in their answers from a nervous dread of the ordeal, but the granting of the document is very largely influenced by the personal experience of the candidate in the practical duties of the engine and boiler room, which must be stated and certified to by the evidence of others.

A general knowledge of the subject of steam engineering is the first requisite to success. A few sample questions are here given to show the ordinary course pursued by examiners to determine the fitness of applicants:

How long have you been employed as an engineer, and where? Are you a mechanic? Where did you learn your trade? Give some idea of the extent of your experience as an engineer? What kind of boilers have you had charge of? Describe a horizontal tubular boiler. Describe a locomotive style boiler. Describe a vertical style boiler. Describe a sectional water tube boiler. How thick is the iron in the shell of your boiler? How thick should it be in the shell of your boiler? How thick are the heads in your boiler? How thick should they be in your boiler? How are the heads fastened to the shell? What is the best way to put heads in a boiler? How is the shell riveted? What size rivets are used? What distance apart are they? How should the shell be riveted? Why do they double rivet some seams? What ones are best double riveted? How is a horizontal boiler braced? How is a locomotive boiler braced? What is the size of and forms of braces generally used? What is the size of your boiler or boilers, length and diameter? How many have you in charge? Name the horse-power. How many tubes are in the boiler? What size are they, and how thick? How long are they? How are they secured? What is the difference between a socket and a stay bolt? What is the tensile strength of Boiler Iron? What is the tensile strength of Boiler Steel? What is mild steel? What is CH No. 1 Iron? What is Flange Iron? What is Hot Short and Cold Short Iron? What is the common dimensions of a Man Hole? What is it for? What are Hand Holes for? Do you open them often? How often? What are Crown Bars and where are they used? How is a Boiler Caulked? What is a Drift Pin?

MECHANICAL STOKERS.

In the back counties of England for many generations before the steam engine was evolved from the brains of Trevethick, Watt and Stephenson, the word "stoke" was used, meaning to "stir the fire." The word was derived from an ancient word, stoke, meaning a stick, stock or post.

To-day there are very many men who are called "stokers," employed principally on locomotive engines, steam vessels, etc., and then there is the "stoke-hole," so-called, in which they do their work.

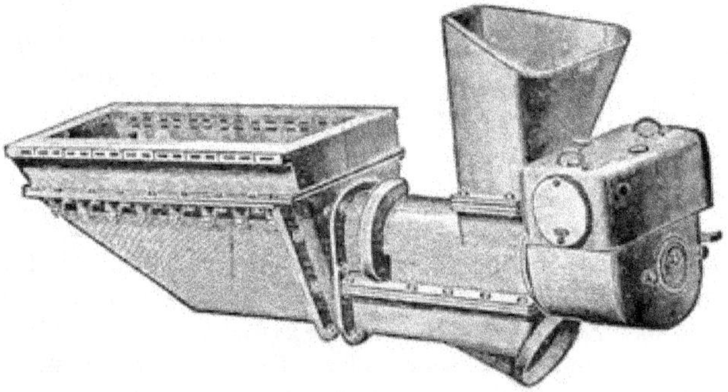

But, now comes the "mechanical stoker," which is well named, as its office is to feed and "stir the fire" by a machine, thus relieving the fireman from much excessively hard toil and allowing the time and energy thus saved to be more profitably used elsewhere. The figure shows a view of the American Stoker which is a device of the most advanced type.

The principal parts of the machine are: 1, the Hopper, which may be filled either by hand shoveling or by elevating and conveying machinery; 2, the Conveyor Screw, which forces the coal, or indeed, any description of fuel, forward to the, 3, Magazine, shown in the figure to the left; 4, a Driving Mechanism, which is a steam motor arranged conveniently in front of the hopper; 5, the Retort, so called from its being the place (above the conveyor) where the coal is distilled into gas.

NOTE.—An illustrated printed description of this machine is issued and sent free upon application by the makers, The American Stoker Co., Washington Life Building, Cor. Broadway and Liberty St., New York.

MECHANICAL STOKERS.

The rate of feeding coal is controlled by the speed of the motor, this being effected by the simple means of throttling the steam in the supply pipe to the motor. The shields covering the motor effectually protect the mechanism from dirt and dust. The motor has a simple reciprocating piston ; its piston rod carries a crosshead, which, by means of suitable connecting links, operates a rocker arm having a pawl mechanism, which in turn actuates the ratchet wheel attached to the conveyor shaft. The stoker is thus entirely self-contained and complete in itself.

A screw conveyor or worm is located in the conveyor pipe and extends the entire length of the magazine. Immediately beneath the conveyor pipe is located the wind box, having an opening beneath the hopper.

At this point is connected the piping for the air supply, furnished at low pressure by a volume blower. The other end of the wind-box opens into the air space between the magazine and outer casing. The upper edge of the magazine is surrounded by tuyeres, or air blocks, these being provided with openings for the discharge of air, inwardly and outwardly.

The stoker rests on the front and rear bearing bars ; the space between the sides of the stoker and side walls is filled with iron plates, termed "dead grates." Steam is carried to the motor by a ¾-inch steam pipe. The exhaust steam from the motor is discharged into the ash pit.

In operation the coal is fed into the hopper, carried by the conveyor into the magazine, which it fills, "overflows" on both sides, and spreads upon the sides of the grates. The coal is fed slowly and continuously, and, approaching the fire in its upward course, it is slowly roasted and coked, and the gases released from it are taken up by the fresh air entering through the tuyeres, which explodes these gases and delivers the coal as coke on the grates above. The continuous feeding gives a breathing motion to this coke bed, thus keeping it open and free for the circulation of air.

It will be noted that in this machine the fuel is introduced from the bottom of the bed of fuel, technically speaking, upon the principle of "underfeeding."

CHEMICAL TERMS

AND EXPLANATIONS RELATING TO FEED WATERS.

Chemistry is a science which investigates *the composition and properties of material substances.*

Nature is composed of elementary elements; knowledge of these bodies, of their mutual combinations, of the forces by which these combinations are brought about, and the laws in accordance with which these forces act, constitute chemistry, and the chemistry of steam engineering largely deals with the foreign bodies contained in the feed water of steam boilers.

Element. In general, the word element is applied to any substance which has as yet never been decomposed into constituents or transmuted to any other substance, and which differs in some essential property from every other known body. The term simple or *undecomposed substance* is often used synonymously with element.

There are about 70 *simple elements,* three quarters of which are to be met with only in minute quantities and are called rare elements. Copper, silver, gold, iron, and sulphur are simple elements—*the metal irridium, for example, is a rare element*—it is the metal which tips the ends of gold pens—it is heavier than gold and much more valuable. Probably there are not two tons of it in existence.

A Re-agent is a chemical used to investigate the qualities of some other chemical—example, hydro chloric acid is a re-agent in finding carbonic acid in lime stone, or carbonate of lime, which when treated by it will give up its free carbonic acid gas, which is the same as the gas in soda water.

An Oxide is any element, such as iron, aluminium, lime, magnesia, etc., combined with oxygen. To be an oxdide *it must pass through the state of oxidation.* Iron after it is rusted is the oxide of iron, etc.

A Carbonate is any element, such as iron, sodium, etc., which forms a union with carbonic acid—the latter is a mixture of carbon and oxygen in the proportion of 1 part of carbon to 2 of oxygen. Carbonic acid, as is well known, does not support combustion and is one of the gases which come from perfect

CHEMICAL TERMS RELATING TO FEED WATER.

combustion. This acid, or what may be better termed a gas, is plentifully distributed by nature and is found principally combined with lime and magnesia, and in this state (*i. e.*, carbonate of lime and carbonate of magnesia) is one of the worst enemies to a boiler.

An Acid is a liquid which contains both hydrogen and oxygen combined with some simple element such as chlorine, sulphur, etc. It will always turn blue litmus red, and has that peculiar taste known as acidity; acids range in their power from the corrosive oil of vitriol to the pleasant picric acid which gives its flavor to fruits.

Alkalies are the opposite to an acid; they are principally potash, soda and ammonia—these combined with carbonic acid form carbonates. Sal-soda is carbonate of soda.

A Chloride is an element combined with hydro chloric acid—common salt is a good example of a chloride—being sodium united with the element chlorine, which is the basis of hydro chloric acid. Chlorides are not abundant in nature but all waters contain traces of them more or less and they are not particularly dangerous to a boiler.

Sulphates are formed by the action of sulphuric acid (commercially known as the oil of vitriol) upon an element, such as sodium, magnesia, etc. The union of sodium and sulphuric acid is the well-known glauber salts—this is nothing more than sulphate of soda; *sulphate of lime is nothing more than gypsum.* Sulphates are dangerous to boilers, if in large quantities *should they give up their free acid*—the action of the latter being to corrode the metal.

Silica is the gritty part of sand—it is also the basis of all fibrous vegetable matter—a familiar example of this is *the ash* which shows in packing, which has been burnt by the heat in steam; by a peculiar chemical treatment silica has been made into soluble glass—a liquid. 65 per cent. of the earth's crust is composed of silica—it is the principal part of rock—pure white sand is silica itself—it is composed of an element called *silicon* combined with the oxygen of the air. Owing to its abundance in nature and its peculiar solubility it is found largely in all waters that come from the earth and is present in all boiler scale.

CHEMICAL TERMS RELATING TO FEED WATER.

In water analysis the term *insoluble matter*, is silica. This is one of the least dangerous of all the impurities that are in feed water.

Magnesia is a fine, light, white powder, having neither taste nor smell, almost insoluble in boiling, but less so in cold water. Magnesia as found in feed water exists in two states, oxide and a carbonate, when in the latter form and free from the traces of iron, tends to give the yellow coloring matter to scale—in R. R. work, yellow scale is called magnesia scale.

Carbonate of Magnesia is somewhat more soluble in cold than in hot water, but still requires to dissolve it 9,000 parts of the latter and 2,493 of former.

Magnesia, in combination with silica, enters largely into the composition of many rocks and minerals, such as soapstone, asbestos, etc.

Lime, whose chemical name is *calcium*, is a white alkaline earthy powder obtained from the native carbonates of lime, such as the different calcerous stones and sea shells, by driving off the carbonic acid in the process of calcination or burning.

Lime is procured on a large scale by burning the stone in furnaces called kilns, either mixed with the fuel or exposed to the heated air and flames that proceed from side fires through the central cavity of the furnace in which the stones are collected.

The calcined stones may retain their original form or crumble in part to powder; if protected from air and moisture they can afterwards be preserved without change.

Soda is a greyish white solid, fusing at a red heat, volatile with difficulty, and having an intense affinity for water, with which it combines with great evolution of heat.

The only reagent which is available for distinguishing its salts from those of the other alkalies is a solution of antimoniate of potash, which gives a white precipitate even in diluted solutions.

Sodium *is the metallic base of soda.* It is silver white with a high lustre; crystallizes in cubes; of the consistence of wax at ordinary temperatures, and completely liquid at 194°, and volatilizes at a bright red heat. It is very generally diffused throughout nature though apparently somewhat less abundantly than potassium in the solid crust of the globe.

Salt, the chloride of sodium, a natural compound of one atom of chloride and one of sodium. It occurs as a rock interstratified with marl, and sandstones, and gypsum, and as an element of salt springs, sea water, and salt water lakes.

The proportions of its elements are 60.4 per cent. of chlorine and 39.6 per cent. of sodium.

In salt made of sea water the salts of magnesia with a little sulphate of lime are the principal impurities.

The above mentioned chemical substances can be classified into two distinct classes, *i.e.*, incrusting and non-incrusting.

Of the incrusting salts, carbonate of magnesia is the most objectionable and any feed water that contains a dozen grains per gallon of magnesia can be expected to have a most injurious effect on the boiler causing corrosion and pitting. Carbonate of lime, while not as bad as the magnesia carbonate, yet has a very destructive action on a boiler and 20 grains per gallon of this is considered bad water. All silicates, oxides of iron, and aluminium, and sulphate of lime are also incrusting. The non-incrusting substances are three, viz., chloride of sodium (common salt), and sulphate and carbonate of soda.

NOTE.

In view of the increasing importance laid upon a knowledge of the chemical formation of feed water, these chapters of Chemical Terms and Analysis of Feed Waters are given to indicate *the direction in which the advanced engineer must push his inquiries.* There are more millions of treasure to be made by properly "treating" the water which enters the steam generators of the world than can be extracted from its gold mines.

An important "point" is to make sure, before adopting any permanent system for purifying the waters of a steam plant, that it is always the same in its ingredients, *i. e.* that the impurities contained in the water are the same at all times.

ANALYSIS OF FEED WATER.

In response to a generous offer made by a leading engineering journal, the following compositions of feed water were ascertained and published. The "Directions" show how the water was forwarded, and the tables the result of careful examination of samples sent from widely separated sections of the country.

Directions.

1. Get a clean gallon jug or bottle and a new cork (or, at all events, a thoroughly clean one).

2. Wash out the vessel two or three times with the same water that is going to be sent in it. This is to make sure that the sample may not be contaminated with any "foreign" ingredient.

3. Tie the cork, after the bottle is filled with the water, with a strong string or wire. Pack the bottle so secure, with hay or straw, sawdust, or newspapers, that it may not knock itself to pieces against the sides of the box.

FROM ARGOS, IND.

	Grains per Gallon.
Silica	1.1096
Oxides of iron and aluminium	.1752
Carbonate of lime	11.9010
Carbonate of magnesia	5.4597
Carbonate of soda	1.1324
Chloride of sodium	.0715
Total solids	19.8494

FROM SIOUX FALLS, S. D.

	Grains per Gallon.
Silica	.8292
Oxides of iron and aluminium	.2452
Carbonate of lime	9.0699
Carbonate of magnesia	5.4376
Chloride of sodium	1.7172
Sulphate of sodium	4.5245
Sulphate of lime	2.6976
Total solids	25.0936

ANALYSIS OF FEED WATER.

FROM LITCHFIELD, ILL.	Grains per Gallon.
Silica	.4711
Oxides of iron and aluminium	.7475
Carbonate of lime	.3800
Carbonate of magnesia	2.2911
Chloride of sodium	8.7543
Sulphate of soda	16.0329
Sulphate of lime	2.8168
Total solids	31.4835

FROM CHELSEA, MASS.	Grains per Gallon.
Silica	.1168
Oxides of iron and aluminium	.6540
Carbonate of lime	34.5260
Carbonate of magnesia	22.8470
Chloride of sodium	63.2041
Sulphate of soda	28.4711
Carbonate of soda	32.2321
Total solids	182.0511

FROM MEMPHIS, TENN.	Grains per Gallon.
Silica	.8292
Oxides of iron and aluminium	.4789
Carbonate of lime	1.8337
Carbonate of magnesia	.9956
Carbonate of soda	1.9792
Total solids	6.1166

FROM PEKIN, ILL.	Grains per Gallon.
Silica	1.0628
Oxides of iron and aluminium	Trace
Carbonate of lime	10.0915
Carbonate of magnesia	5.8224
Chloride of sodium	Trace
Sulphate of soda	1,2456
Total solids	18.6471

FROM TIFFIN, OHIO.	Grains per Gallon.
Silica	.5256
Oxides of iron and aluminium	.2336
Carbonate of lime	12.6144
Carbonate of magnesia	10.2652
Carbonate of soda	2.4137
Sulphate of soda	6.8296
Chloride of sodium	1.0484
Total solids	33.9305

CORROSION AND INCRUSTATION OF STEAM BOILERS.

No more perplexing question presents itself to the engineer and steam user than the one to be inferred from the above heading. Enormous losses of money, danger to life and property and the loss of position and the reputation of the engineer are involved in it. How to avoid these actual evils is of the first importance in steam economy. The subject at first sight seems to the average student a difficult one to master, but like all other matters pertaining to mechanics, investigation that is backed with reason, will show that much that appears obscure is really very plain indeed; this is because nature, even down to the sediment remaining in a boiler after the conversion of water into steam, operates in its formation with infinite exactness and along well known lines.

Question.—What is corrosion ?

Answer.—*Corrosion is simply rusting* or the wasting away of the surfaces of metals, for particulars of which see page 126.

Question.—What is Incrustation ?

Answer.—*Incrustation means* simply *a coating over.*

Water, on becoming steam, is separated from the impurities which it may have contained, and these form sediment and incrustation.

Boilers corrode *on the outside as well as within,* and to a great extent unless carefully cleaned and painted ; but it is the damage caused by "hard" and accidulated water within the boiler that is to be principally guarded against.

An extreme example of incrustation has been described in that of a locomotive type of a stationary boiler. Its dimensions were : seventy-two inches in diameter, twenty-two feet long, with 153 three-inch tubes ; shell, three-eighths ; head, three-eighths, and made of iron. The scale against the back head was nearly two inches thick and completely filled the space between the tubes, so that circulation was impossible, the only wonder being that the boiler did not give out sooner than it finally did. The scale was even with the top row of tubes, the

CORROSION AND INCRUSTATION OF STEAM BOILERS.

only part of the boiler generating steam being the fire box and the upper row of tubes, the others acting simply as smoke conduits. There was certainly a great loss of fuel, quite fifty per cent. Had it been a horizontal boiler it would have burned out before the scale became so heavy.

In the above instance, the loss in fuel is estimated at one-half. Careful experiment has proved an average loss of fuel as follows:

1-16	inch of	scale	causes	a loss of	13	per cent	of fuel.
1-4	"	"	"	"	38	"	"
1-2	"	"	"	"	60	"	"

It must be remembered that dry steam, as it is used through the engine or for other purposes, *carries away none of the impurities* which pass with the water into the boiler; hence, in a battery of boilers burning, say, 20 tons of coal per day and evaporating 10 lbs. of water to a pound of coal, there is a body of water going through them every day of 200 tons. Multiply this by 300 days for a year=60,000 tons, and it will be seen how very great is the problem of keeping the interior of the boilers free from scale and deposit.

Chemically pure water is that which has no impurities, and may be described as colorless, tasteless, without smell, transparent, and in a very slight degree compressible, and, were a quantity evaporated from a perfectly clean vessel, there would be no solid matter remaining.

But, strangely, investigation has proved that water of this purity rapidly corrodes iron, and attacks even pure iron and steel more readily than "hard" water does, and sometimes gives a great deal of trouble where the metal is not homogeneous. Marine boilers would be rapidly ruined by pure distilled water if not previously "scaled" about 1-32 of an inch.

Water is formed by the union of two gases—oxygen and hydrogen. These two are *simple bodies,* formed by the Creator in the beginning, which are found *in combination* in thousands of different forms. Both when alone are invisible. Take one volume of oxygen and mix it with two volumes of hydrogen and they will chemically unite and form water. This is by measure. *By weight* water is composed of 88.9 of oxygen to 11.1 of hydrogen = 100 parts. See pages 229, 230 for further information.

It is an important point to remember that when water is expanded about 1,700 times into steam, it is simply expanded water, as ice is hardened water, *i.e.*, in expanding into steam the two constituent gases do not separate. Hence, in dealing with the impurities inside the boiler, it is to be observed that in no sense do they change the essential nature of water itself. The impurities are simply *foreign bodies*, which have no legitimate place in the boiler, and are to be expelled as dangerous foes. As a general principle, it may be stated that it is more profitable to soften and filter the water used in boilers than to trust to blowing out or dissolving the sediment and scale that will be otherwise formed, for observations show that "anti-incrustators" containing organic matter help rather than hinder incrustations, and are therefore to be avoided. For the remedy of foul water there are numerous contrivances to prevent it from entering the boiler, which is far better than trying to extract the sediment after it is there, though there are many ingenious methods for doing that also, some of which will be detailed hereafter.

PRELIMINARY PRECIPITATION OF WATER.

A good method of avoiding incrustations in steam boilers is evidently a preliminary purification of the feed-water, provided it can be done by means sufficiently simple. This is a problem which it is claimed has been solved by M. Dehne of Halle, by means of an arrangement which we will herewith describe. The fresh water, which is taken up by a feed pump, is sent into a heater where it is raised to a temperature that will be favorable to chemical reaction. It then passes into a mixer where it encounters certain reacting agents which have been pumped in there by a pump of special design. These reacting agents are composed of a mixture of carbonate of soda and of caustic soda, the carbonate of soda serving to precipitate the sulphate of lime contained in the feed-water, while the caustic

CORROSION AND INCRUSTATION OF STEAM BOILERS.

soda precipitates the carbonate of lime and the magnesia. The relative dimensions between the special pump and the feed pump are calculated in such a way that the proportions of carbonate of soda and caustic soda in the mixture have always a certain relation to the amount of lime and magnesia to be precipitated. The water of the mixture is frequently very much disturbed by the precipitations which are formed, and passes into a filter where all the matters that are held in suspension are retained. It then goes into the boiler. In cases where the feed-water is taken from a tank, the heater, the mixer, and filter are put in the suction pipe of the feed pump, but if, as often happens, the water is already under pressure and will pass directly through the three, the feed pump will take the water directly from the filter and pump it directly into the boiler.

A PRECIPITATOR FOR SEA WATER.

It is quite possible to prepare sea water in such a way as to practically prevent any serious deposit forming from it.

The process employed is to add to the sea water a known quantity of precipitator powder consisting chiefly of soda ash, and having done this in a closed vessel, to heat the mixture by blowing into it waste steam, until a pressure of from 5lb. to 10lbs. is created; under these circumstances practically all the magnesium and calcium salts separate from the water and are easily got rid of by filtering it under pressure into the hot-well.

A precipitator 6 ft. 4 in. high and 3 ft. in diameter, holds a ton of water, and the time taken, from the first running the sea water in, to its delivery into the hot-well, need not exceed 1 hour and 15 minutes, so that in practice, giving plenty of time between the makes, it would be perfectly easy to prepare 8 to 12 tons in the 24 hours with a small precipitator of the size named. The prepared water has a density of 1-32nd, and may with safety be evaporated until its density is 5-32nds, the salts present not crystallizing out until a density of from 6-32nds to 7-32nds is reached.

In preparing sea water in the way proposed, every precaution must be taken to add slightly less of the precipitant than is necessary to entirely throw down the calcium and magnesium

A PRECIPITATOR FOR SEA WATER.

salts, as it is manifestly impossible in practice to guard against small quantities of sea water finding way into the boiler either from leaky condensers or else being fed in by the engineer during some emergency, and if under these conditions any excess of the precipitant were present in the boiler, a bulky precipitate would be thrown down and cause trouble, although it would not bind into a solid scale.

Briefly recapitulated the means which are best adapted for preventing the formation of the dangerous organic and oily deposits considered are:

I. Filtration of condensed water through a coke column.

II. Free use of the scum cocks.

III. The use of water of considerable density rather than of fresh water.

IV. The use of pure mineral oil lubricants in the smallest possible quantity.

SCALE DEPOSITED IN MARINE BOILERS.

The analysis given below may be looked upon as typical of the incrustation formed by fresh water, brackish water and sea water respectively in marine boilers:

Constituent	River.	Brackish.	Sea.
Calcic carbonate	75.85	43.65	0.97
" sulphate	3.68	34.78	85.53
Magnesic hydrate	2.56	4.34	3.39
Sodic chloride	0.45	0.56	2.79
Silica	7.66	7.52	1.10
Oxides of iron and alumina	2.96	3.44	0.32
Organic matter	3.64	1.55	trace
Moisture	3.20	4.16	5.90
	100.00	100.00	100.00

From this it is evident we may look upon the incrustation from fresh water as consisting of impure calcic carbonate, whilst that from sea water is impure calcic sulphate, the brackish water from the mouths of rivers yielding, as might be expected, an incrustation in which both these compounds are present in nearly equal quantities.

The importance of these differences in the deposit formed is very great, as it enables the shipowner to arrive at the conclusion as to the treatment that the boilers have received during the voyage, by examination and analysis of the scale that those

SCALE DEPOSITED IN MARINE BOILERS.

boilers contain. Taking, for instance, the case of a ship which uses fresh water both for filling and make up, it is manifest that on her return to port the scale should be very slight and should consist mainly of calcic carbonate, whilst, if the scale exceeds 1-16 in., and shows a preponderance of calcic sulphate, it is manifest that such scale could only have been formed by sea water, either leaking in from faulty condensers or being deliberately fed into the boilers.

With the introduction of high pressure steam a new and dangerous form of deposit has added to the trouble of the marine engineer; having entered the boiler, the minute globules of oil, if in great quantity, coalesce to form an oily scum on the surface of the water, or if present in smaller quantities, remain as separate drops; but show no tendency to sink, as they are lighter than water.

Slowly, however, they come in contact with small particles of other solids separating from the water and sticking to them, they gradually coat the particles with a covering of oil, which in time enables the particles to cling together or to the surfaces which they come in contact with. These solid particles of calcic carbonate, calcic sulphate, etc., are heavier than the water, and, as the oil becomes more and more loaded with them, a point is reached at which they have the same specific gravity as the water, and then the particles rise and fall with the convection currents which are going on in the water, and stick to any surface with which they come in contact, in this way depositing themselves, not as in common boiler incrustation, where they are chiefly on the upper surfaces, but quite as much on the under sides of the tubes as on top.

The deposit so formed is a wonderful non-conductor of heat, and also from its oily surface tends to prevent intimate contact between itself and the water. On the crown of the furnaces this soon leads to overheating of the plates, and the deposit begins to decompose by heat, the lower layer in contact with the hot plates giving off various gases which blow the greasy layer, ordinarily only 1-64 inch in thickness, up to a spongy leathery mass often 1-3 inch thick, which, because of its porosity is an even better non-conductor of heat than before, and the plate becomes heated to redness.

When water attains a temperature, as it does under increasing pressure, ranging from 175° to about 420° Fahr., all carbonates, sulphates and chlorides are deposited in the following order:

First. Carbonate of lime at 176° and 248° Fahr.

Second. Sulphate of lime at 248° and 420°.

Third. Magnesia, or chlorides of magnesium, at 324° and 364°.

It is to take advantage of this fact that mechanically arranged jets, sprinklers and long perforated pipes are introduced into the interior of the boiler; these tend to scatter the depositing impurities and also to bring the feed water more quickly to the highest heat possible.

With regard to the oxide of iron or iron salts in solution, these can best be treated with small quantities of lime. By adding re-agents, they set up chemical changes, which result in precipitation, which give the water a milky appearance; they divide into particles, and ultimately settle, leaving the water pure and bright. The mechanical treatment on a limited scale would be easy, a settling tank sufficing; but this becomes a different matter when large quantities have to be dealt with.

ANALYSIS OF AVERAGE BOILER SCALE.

	Parts per 100 parts of deposit.	
Silica	.042	parts.
Oxides of iron and aluminium	.044	"
Carbonate of lime	30.780	"
Carbonate of magnesia	51.733	"
Sulphate of soda	Trace	"
Chloride of sodium	Trace	"
Carbonate of soda	9.341	"
Organic matter	8.060	"
Total solids	100.	Parts

The percentage only of each ingredient the scale is composed of is given, as it cannot be told how much water was evaporated to leave this amount of solid matter.

A LOCOMOTIVE-BOILER COMPOUND.

The lines of a certain great R. R. traverse a country where the water is very hard and they are compelled to resort to some method of precipitating the lime that is held in solution. After many tests and experiments they have made a compound and use it as follows: in a barrel of water of a capacity of fifty gallons they put 21 lbs. of carbonate of soda, or best white soda ash of commerce, and 35 lbs. of white caustic soda. The cost, per gallon, is about 2½ cents.

The compound is carried in this concentrated form, in calomine cans on the tender of each locomotive. A certain amount, according to the necessities of the case, is poured into the tender at the water tank at each filling. This amount is determined by analysis, and varies all the way from two to fifteen pints to two thousand gallons of water. The precipitating power of this compound may be taken roughly at ⅔ of a pound of the carbonate of lime, or equivalent amount of other material, per pint of the compound. On their western lines where they are dealing with alkali waters and those containing sulphates, the company use merely 60 pounds of soda ash to a barrel of water. When the water is pumped into the boiler the heat completes the precipitation and aggregation of the particles, and this does away with all trouble of the tenders or injector tubes clogging up.

The case is an interesting one to stationary engineers, because where the water is pumped into the boiler from tanks the same compound can be used, provided the water contains the proper constituents to be precipitated by it; and where the water is taken from city water mains, it would be a simple matter to devise an apparatus to admit the compound to the feed pipes.

"Points" Relating to the Scaling of Steam Boilers.

The peculiarity about the sulphate of lime is that *the colder the water the more of it will be held in solution.* Water of ordinary temperature may hold as high as 7 per cent. of lime sulphate in solution, but when the temperature of the water is raised to the boiling point a portion of it is precipitated, leaving about .5 of one per cent. still in solution. Then as the

POINTS RELATING TO THE SCALING OF STEAM BOILERS.

temperature of the water is raised, still more of the substance is precipitated and this continues until a guage pressure of 41 pounds has been reached which gives a temperature of about 200 degrees; at this point all the sulphate of lime has been precipitated. Many other scale forming substances act in a similar manner. This shows quite plainly that any temperature that can be produced by the use of exhaust steam would not be sufficient to cause the precipitation of all the substances which might be contained in the water.

That boiler incrustations are the immediate causes of the majority of steam boiler explosions is no longer a doubtable question.

Nearly all foreign matter held in solution in water, on first becoming separated by boiling, *rises to the top in the form of what is commonly called scum*, in which condition much of it may be removed by the surface blow-off. If not removed, however, the heavier particles will be attracted to each other until they have become sufficiently dense to fall to the bottom, where they will be deposited in the form of scale, covering the whole internal surface of the boiler below the water line, with a more or less perfect non-conductor of heat.

It is recorded that the engineer of the French ocean steamer *St. Laurent* omitted to remove a bar of zinc when repairing and cleaning out his boilers. On opening the boilers at the end of the voyage to his great surprise he found that the zinc had disappeared, but his boilers were entirely free from scale and the boiler plates not injured in the least.

It has been recently determined by some German experimenters that sugar effects a strong action upon boilers. It has an acid reaction upon the iron which dissolves it with a disengagement of hydrogen. The amount of damage done increases with the amount of sugar in the water. These results are worthy of note in sugar refineries and places where sugar sometimes finds its way into the boilers by means of the water supplied. The experimenters in question also find that zinc is strongly attacked by sugar ; copper, tin, lead and aluminium are not attacked.

POINTS RELATING TO THE SCALING OF STEAM BOILERS.

Two reasons, relating to incrustations, for not blowing out a boiler while under steam pressure may be given as follows: One is, that the foreign matter floating on top of the water will be deposited on the shell of the boiler as the water gradually subsides, and, second, the heated walls of the furnace will communicate a sufficiently high temperature to the boiler to dry and flake the sediment that would otherwise remain in the boiler in the shape of mud, which could easily be washed out, were it not for the baking process.

Bark, such as is used by tanners, has an excellent effect on boiler incrustations. It may be used as follows: Throw into the tank or reservoir from which the boilers are fed a quantity of bark in the piece, in sufficient quantity to turn the water to a light brown color. Repeat this operation every month at least, using only half the quantity after the first month. Add a very small quantity of the muriate of ammonia, about one pound for every 2,000 gallons of water used. This will have the effect of softening as well as disintegrating *the carbonate of lime* and other impurities deposited by the action of evaporation.

NOTE.—Care must be exercised in keeping the bark, as it becomes broken up, from the pump valves and blow-off valves. This may be accomplished by *throwing it into the reservoir confined in a sack.*

Among the best samples of boiler compounds ever sent to the laboratory for analysis was found to be composed of:

	Pounds.
Sal soda	40
Catichu	5
Sal ammoniac	5

This solution was formerly sold at a good round figure, but since its nature became more generally known, it is not found in market, but is largely used, consumers putting it up in lots sufficient to last a year or so at a time.

The above is strongly recommended by those who have used it, *one pound of the mixture being added to each barrel of water used,* but after the scale is once thoroughly removed from the

POINTS RELATING TO THE SCALING OF STEAM BOILERS.

boiler, the use of sal soda alone is all that is necessary. By the use of ten pounds per week a boiler 26 feet long and 40 inches in diameter in one of the iron mills of New Albany, Ind., has been kept clean of scale equal to a new boiler.

There are other evils sometimes inherent in hard waters over and above the mere production of a crust. Some waters contain a great deal of soluble magnesia salts, together with common salt. When this is the case there is a great chance of corrosion, for the former is acted on by steam at high pressure in such a way that muriatic acid fumes are produced, which seriously corrodes the boiler, and, what is far worse, passes with the steam into the engine, and produces corrosion in the cylinders and other delicate fittings into contact with which the steam passes. All this can, however, be obviated by the removal of the magnesia from the water.

There has not been, and never can be, made a mechanical device which will precipitate all the ingredients contained in a water taken from a natural source of supply, and if it were possible to do so it would be the most ruinous thing one could do for the boilers, as water is the greatest *solvent* known to chemistry, and its nature is to hold in solution and be impregnated with the different elements it comes in contact with, to a certain per cent., and if its lime, magnesia, and the mineral salts are taken away, and the pure water is pumped into the boilers, it will take up the iron, causing pitting and grooving of the boilers. It is better to let nature take its course, to a certain extent, and neutralize what little mineral deposit forms in the boilers with as small an amount of vegetable matter as possible.

It is well to note that different waters require different treatment; what will be of benefit in one instance will be of no value whatever in a different water, many of the "compounds" sold to prevent and remove scale will certainly destroy a boiler if they are used persistently, because they are composed of the exact opposite chemicals which should be used; as an example it is stated that at one establishment one thousand dollars were expended annually for a mixture which it was said resulted in the reduction of the life and usefulness of the boilers of 50 per cent.

ENGINEERS' TESTS

FOR IMPURITIES IN FEED WATER.

Much expense can be saved in fuel and boiler repairs by a little preliminary expenditure of money in securing a supply of good water for the steam boilers of a new establishment. Well water is nearly always inferior to the running water of streams; water from mines is especially hurtful, containing, as they do, large quantities of free sulphuric acid. Wells along the sea shore or on the banks of rivers affected by the tides, are likely to be saturated with chloride of magnesium. It is in determining these points that these ready tests of feed water are most useful.

A thorough and really scientific analysis of feed water is a costly and tedious process, but *a simple and perhaps sufficiently accurate test* may be made as follows: take a large (or tall) clear glass vessel and fill it with the water to be tested: add a few drops of water of ammonia, until the water is distinctly alkaline: next add a little phosphate of soda; the action of this is to change the lime, magnesia, etc., into phosphates, in which form they are deposited in the bottom of the glass. The amount of the matter thus collected gives a crude idea of the relative quality of sediment and scale-making material in the water.

Water turning *blue litmus paper red,* before boiling, contains carbonic acid, and if the blue color *can be restored by heating,* the water contains carbonic acid. Litmus paper is sold by druggists.

If the water has a foul odor, giving a black precipitate with acetate of lead, it is sulphurous.

An experiment may be tried by dissolving common white or other pure soap in a glass of water, and then stirring into the glasses of water to be tested a few teaspoonsful of the solution; the matter which will be deposited will show the comparative amount of the scale-making material contained in the feed water.

ENGINEERS' TESTS FOR IMPURITIES IN FEED WATER.

In order to ascertain the proportion of soda to the feed water the following method is recommended :

1. Add $\frac{1}{16}$th part of an ounce of the soda to a gallon of the feed water *and boil it.* 2. When the sediment thrown down by the boiling has settled to the bottom of the kettle, pour the clear water off, and 3, add ½ drachm of soda. Now, if the water remains clear, the soda, which was first put in, has removed the lime, but if it becomes muddy, the second addition of soda is necessary.

In this way a sufficiently accurate estimate of the quantity of soda required to eliminate the impurities of the feed water can be made and the due proportion added to the feed water.

By exercising a little judgment, the use of pure chemicals, with well cleaned vessels, test tubes, etc., the following reagents will determine the character of the most important elements which injure the iron surfaces of a steam boiler.

Carbonic acid is indicated by byrata water.
Sulphates are indicated by chloride of barium.
Chlorides are indicated by nitrate of silver.
Lime salts are indicated by oxalate of ammonia.
Organic matter is indicated by chloride of mercury.

The "base" of the better class of the various patented boiler compounds is tannin (whence tannic acid) and some form of alkali, and if the compounds were to be deprived of these two elements they would be absolutely worthless.

Where they contain, as some certainly do, sal-ammoniac, muriatic, hydrochloric and sulphuric acids, they cannot but act as boiler destroying agents.

Tannin or tannic acid is the principal ingredient used in preparing leather. It is found in a great variety of plants—sassafras root has it in large proportion, the gall nut and the bark of various trees, especially the oak produce it.

It is the presence of this acid that gives their only value to very many "compounds," tan bark, gum catechu (which sometimes contains one-half part of tannic acid), etc. The acid seems to have but little effect where large quantities of sulphate of lime are present, but in waters where carbonate of lime predominates its detersive qualities are more marked.

The records of the Patent Office shows that one boiler compound *contains* 23 *per cent. of catechu,* and others 60, 81, 5, respectively, by which may be inferred the large quantity of this agent, which has been sold in combination with other chemicals, principally soda.

NOTE.

While the product of water steeped in clean tan bark may be favorable in its action upon boiler incrustation, *it has been found to be very unsafe, in practice, to use the "tan liquor" taken from the vats.* The danger arises from the fact that sometimes during the process of tanning leather, the required acidity cannot be produced by natural fermentation when sulphuric acid is added, in order to bring the liquor to its required strength—in due course, this corrosive substance acts injuriously on the boiler.

USE OF PETROLEUM OIL IN BOILERS.

The use of crude (unrefined) mineral oil in steam boilers is attended by risks caused by impurities and foreign substances mixed with it. These are likely to combine with the earthy matter in the water and tend to form instead of preventing scale ; the tar and wax contained in crude petroleum combine with the sediment in steam boilers, and the paste prevents the water from reaching and protecting the plates. This is true particularly in shell boilers which have flat surfaces over the fire. Refined mineral oil has none of these disadvantages.

Kerosene oil has all the advantages to be derived from the use of crude petroleum and the above objections quite removed.

In one system of the application of steam the use of kerosene and petroleum cannot be recommended : that is *when live steam is used for cooking purposes,* the odor from the oil will impregnate the meat and other products designed for food consumption.

KEROSENE OIL IN BOILERS.

Under certain conditions, and with care and judgment, the use of refined petroleum has been found to be of great advantage in removing and preventing scaling in steam boilers.

There is no well authenticated case where a systematic, regular and uniform feed of pure kerosene oil to a steam boiler has failed to operate beneficially upon the scale formation.

The best results are obtained by the use of the oil *under the same arrangement that cylinder oil is fed to an engine.* The kerosene is sometimes introduced through a one-fourth inch branch to the suction pipe of the feed pump, leading to the vessel containing the oil, so that any quantity, large or small, can be put into the boiler simultaneously with the usual feed. The drawback to this arrangement is that when the feed water heater has to be cleaned, a gallon or more of the oil is often lost, which together with a very unpleasant odor, when used in this manner, tends to condemn its use. *But when piped between the boiler and heater,* these objections cease. We present an arrangement which is illustrated by cut on page 157.

This is nothing more than a storage system with sight feed, by use of which the oil can be fed drop by drop as desired—for each drop of water entering the reservoir a drop of oil is forced down the small ¼ in pipe, up the glass tube and on into the boiler.

In piping it is necessary to have the water or larger pipe (½in.) attached through the lower plug as shown in cut, and the oil, as shown, going through the smaller or ¼ in. pipe—i. e., the oil pipe must, under all circumstances, be the smaller of the two.

In the figure is shown a piece of 6 in. gaspipe, about a foot in length, plugged at each end; the top plug has one opening, for an inch nipple "a" with top. This opening is to be used in filling the reservoir with oil. The botton plug has two holes, one for the ½ in. water pipe, and the second for a small pet cock "B," to let the water out, whenever it is necessary to refill the tank with kerosene. The water guage connection is

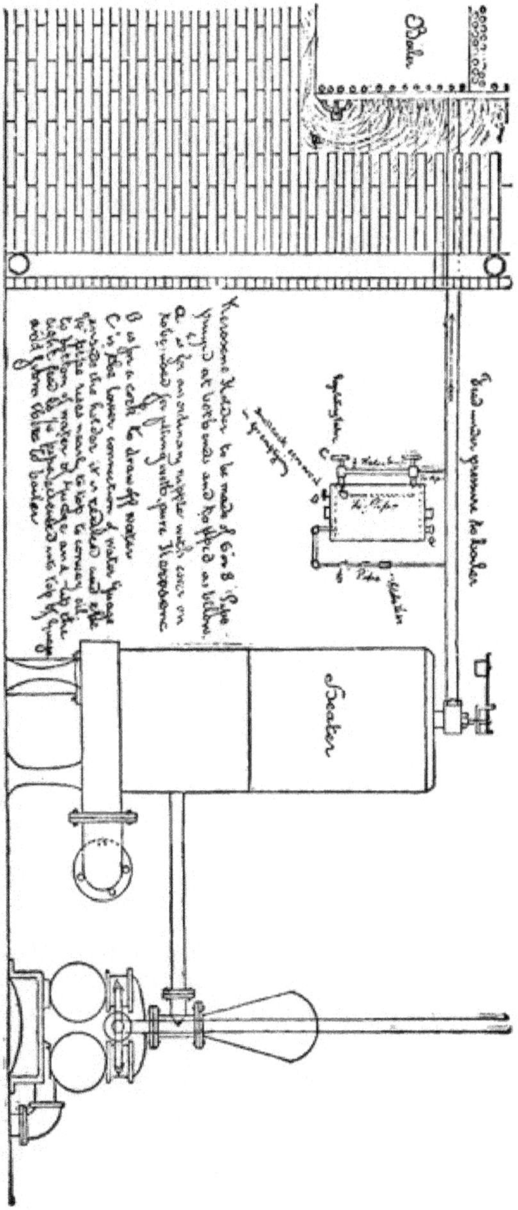

METHOD OF FEEDING KEROSENE OIL TO BOILER—Fig. 69.

DEVICE FOR USING KEROSENE OIL.

the ordinary, cheap brass fixture, with boxes, nipples, etc., used in boilers, with gasket of rubber bottom and top of the glass. The glass plainly exhibits the depth of water and oil in the reservoir as well as the feed of minute drops of oil as they speed on their beneficent mission softening the injurious scale. There are the usual 2 valves on the water glass; by opening the lower one more or less, the amount of oil used can be regulated to a nicety. The valves can be used to entirely cut off the apparatus at any time desired.

NOTE.—Should the end of the screw connection inside the holder which each one of these valves control, not be ¼ inch, a reduced elbow should be used, as ¼ in. pipe will give the best satisfaction when used as a stand pipe inside the reservoir.

The quantity oil to be fed to a boiler is very largely to be determined by experiment commencing with a minimum and increasing the amount as found necessary to keep down the scale formation. The use of 2 qts. of the oil per week has been found to be sufficient for a boiler 4 feet in diameter and 12 feet long, and three quarts per week on boilers 5 feet in diameter. This quantity may be regarded as the smallest advisable to use and from that up to 1 to 2 gallons per diem in boilers, say of 125 horse power, when pushed to their capacity in evaporating water.

The result of careful experiments justifies the use of kerosene, the scale being less than in four years' previous experience, and a large portion of the boiler showing the clean black steel, in as apparently good condition as when new.

Despite the small quantity of kerosene used in the boilers in this case, the odor was perceptible by opening an air valve to any steam radiator in any of the buildings. When as much as a gallon per week was used, the odor was very strong, but with one-half that amount it was hardly perceptible, and only to be noticed when an air valve had been open a long time. And since commencing to use the oil a much greater deposit of rust scales than usual has been found in the various steam traps in the buildings, indicating that the oil is also exerting a cleansing influence on the pipes of the whole system.

DEVICE FOR USING KEROSENE OIL.

NOTE.—Provision must be made for the removal of the scale as it drops from the internal surfaces of the boiler, as at times many bushels of it have been deposited directly over the furnace; hence, if a boiler is known to be badly incrusted, the kerosene should not be put in the first time more than three days before it is intended to wash the boiler.

NOTE 2.—The safety valve should be opened to allow the escape of the gas arising from the kerosene before cleaning out the boiler, where a lighted lamp or candle is used, as it must necessarily be—indeed this is a precaution which ought always to be observed in all cases, viz., properly to ventilate boilers, heaters, and tanks of all descriptions before entering them with lighted lamps and torches. While these gases are not likely to cause an explosion, they burn quite rapidly and should be promptly removed without giving opportunity for an accident.

The accumulation of gas is not confined to the use of kerosene oil for the prevention of scale in steam boilers, but is also found in flour mills, confectioners', conduits for electric wires, brewers' vats, etc. So, with common sense precautions, no extra risk is run in using kerosene oil in steam boilers.

MECHANICAL BOILER CLEANERS.

Owing to the fact (1) that nearly, if not quite all, the impurities which exist in feed water are set free by the high temperature attained under pressure; (2) that these impurities are left in the boiler by the constant use of the steam, there follows the result that the water remaining is more and more impregnated with the residuum composed of the foreign matters which (the water removed) constitutes mud, scale, etc.

The custom has been and is now to regularly "blow off" one or two guages of this water once or twice per day replacing it with fresh water of less density; that this is a very imperfect method for removing the foreign matter is readily allowed, besides wasting absolutely all the units of heat contained in the water blown off.

Now, within the boiler while in use, under the operation of the fierce heat of the furnaces, are constant changes in the

MECHANICAL BOILER CLEANERS.

position of the water caused by the boiling, by the withdrawal of the steam and by the constant effort of the hot water to rise and the cold water to fall. The water thus keeps in circulation everything within the boiler, including the sediment, *except in places where the water is from any cause without motion.* In these quiet nooks there is a constant depositing of the elsewhere active foreign matters contained in the water, which deposits, in the form of mud and scale, left undisturbed, causes loss and danger.

It is in taking advantage of these facts, and of the principles of the circulation of hot and cold water, that mechanical boiler cleaners are brought into successful use.

These devices for the stilling of the water and collection of the sediment are made in various forms and all sizes and capacities, and are located at the sides or back of the boiler setting and even on top of the boiler. There is a system where pipes in a coil are fixed in the sides of the furnace and exposed to its greatest heat, and which, owing to their enlarged area, act as most efficient reservoirs. In all these devices there is an *upflow pipe* connected with the lower and coolest water, and a *return pipe* connecting with the top of the water where it is hottest. This arrangement assures a constant current which is more or less rapid according to the intensity of the fire and which keeps up as long as the firing is done. Where this current passes through the reservoir, the enlarged area and comparative quiet is favorable for the deposit of the sediment and in practical experience it does deposit nearly all of it. The collection of the impurities is helped by a *funnel-shaped appliance* placed at the opening of the upflow pipe, which, aided by the rapid flow of the hot water, carries the floating scum towards it into the reservoir. Attached to the reservoir is the blow-off pipe through which the deposited matter is removed as often as necessary.

The use of these mechanical cleaners is readily understood: (1) they provide a place of accumulation for the sediment; (2) they save the necessity of opening the boilers to remove by hand, the refuse of the boiler; (3) save fuel by avoiding the necessity of frequent blowing off one or two guages of water, and (4) by the preventing the formation of scale with its attendant evils.

SCUMMING APPARATUS.

In addition to the bottom blow-out apparatus every boiler should be provided with means for blowing out water from the surface in order to remove the fine particles of foreign matter floating there, which afterward settle and consolidate as scale on the heating surfaces.

It consists, in its simplest form, of a pan, or a conical scoop,

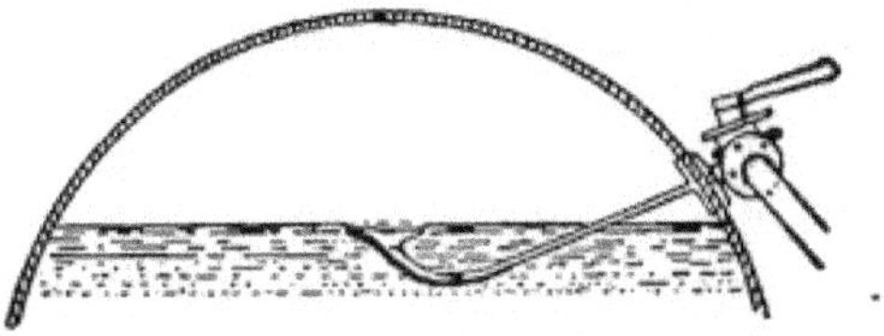

Scum Cock

Fig. 70.

near the surface of the water, but below it, connected with a pipe passing through the boiler-shell, on which is a cock, or valve, for regulating the escape of the water laden with the impurities deposited in the pan. There are patented apparatus for this purpose which are well designed and easily fitted to a boiler.

The office of the surface blow-off, illustrated in Fig. 70, is to remove the foreign matter which is precipitated from its solution in the water.

A surface blow-off used occasionally will remove the greater portion of this scum and keep the boilers reasonably free from scale and mud. Where dirty or muddy water is fed into the boilers the surface blow-off is one of the cheapest and most efficient means for keeping the boiler clean. The efficiency of the surface blow-off is not so great as that of some of the mechanical boiler-cleaners, as by their use it is not required that any hot water shall be wasted, and this is the greatest objection to the surface blow-off, as in the hands of some people a large amount of boiling water is wasted each time it is used. But both of these arrangements are virtually skimmers, as they remove the precipitated mineral and vegetable matter from the surface of the water in the boiler. One does it by blowing out

SCUMMING APPARATUS.

the scum and some water at the same time, while the mechanical boiler-cleaner removes the scum, but returns the water to the boiler.

There are several efficient ways of arranging a surface blow-off. The principal part of the blow-off is a pan or perforated pipe placed horizontally at the water level having a pipe leading outside the boiler to any convenient place where the scum may be blown. When a perforated pipe is used the action is to force the scum from the top of the water during the time the valve is open, and blow it through the pipe. In using an apparatus of this kind it should be blown often, but only for a moment at a time, as all the scum near the pipe is removed immediately, and to keep the valve open longer than necessary to remove the scum near the pipe would allow the escape of clean water or steam which would be wasteful. If a pan is used and is fastened so that the top is secured at the ordinary water level, as shown in Fig. 70, the blow-off pipe leading from near the bottom of the pan, it will be more efficient than the perforated pipe arrangement as it will not require to be used so often, and the waste of water and steam will not be so great. The pan, by producing an eddy in the water, causes all the scum to gather over the top, and as the water is quiet there it will gradually settle into the pan, where it will remain as mud. When the blow-off valve is opened the greater part of the mud which is gathered is blown out, and but very little water is carried with it.

USE OF ZINC IN MARINE BOILERS.

Zinc has been used in marine boilers for many years, but it was not until the publication in 1880 of the report of the Admiralty committee that the use of zinc became general. It has been used in various ways: 1.—Virgin spelter, as imported in oblong slabs of various sizes. 2.—Cast, or remelted zinc. 3.—Cast zinc buttons, generally made from virgin spelter or new clean zinc trimmings. 4.—Zinc spheres. 5.—Rolled zinc blocks, generally 12 inches by 6 inches, and thicknesses varying from ½ inch to 1½ inch, generally with a 13-16 inch hole in the centre.

USE OF ZINC IN MARINE BOILERS

It is desirable that close-grained zinc of uniform structure and free from impurities should be used, and rolled zinc appears to meet this want. The wear is entirely confined to the surface. It does not appear to become distorted or broken up. On the contrary, it gradually wastes away till only a slight shred, a sort of skeleton frame work, remains to indicate what it has been.

The primary object in the use of zinc in boilers is the prevention of corrosion, but it has also some effect in reducing the amount of incrustation, and rendering it softer and less adherent.

Table

Showing Amount of Sediment collecting in a steam boiler when evaporating 6,000 gallons per week, of 58,318 grains each.

When a gallon of feed water evaporated to dryness at 212 degrees Fahrenheit, leaves of solid matter in grains.	The amount of solid matter collecting in boiler per week will be:		When a gallon of feed water, evaporated to dryness at 212 degrees Fahrenheit, leaves of solid matter in grains:	The amount of solid matter collecting in boiler per week will be:	
Grains.	Pounds.	Ounces.	Grains.	Pounds.	Ounces.
1		13.714	55	47	2.285
2	1	11.428	60	51	6.857
3	2	9.143	65	55	11.428
4	3	6.857	70	60	
5	4	4.571	75	64	4.571
6	5	2.285	80	68	9.143
7	6		85	72	13.714
8	6	13.714	90	77	2.285
9	7	11.428	95	81	6.857
10	8	9.142	100	85	11.428
15	12	13.713	110	94	4.571
20	17	2.284	120	102	13.714
25	21	6.855	130	111	6.857
30	25	11.426	140	120	
35	30		150	128	9.142
40	34	4.571	160	137	2.285
45	38	9.143	170	145	11.428
50	42	13.714	180	154	4.571

BOILER FIXTURES AND BELONGINGS.

A boiler is not complete without certain fixtures. There must be a feed-pump or injector, with a supply-pipe, feed-valve, safety feed-valve, and check-valve, in order to supply water properly to the boiler; gauge-cocks, a glass water-gauge, a blow-pipe, with its valve, to reduce the height of the water in the boiler, or to empty it entirely; a safety-valve to allow the steam to escape from the boiler when it exceeds a fixed pressure; a scumming apparatus to remove the foreign matters from the water as much as possible; a steam-pipe to convey the steam to the place where it is wanted; man-holes and hand-holes, with their covers and guards, for examination and cleaning; a non-corrosive steam-gauge, to accurately indicate at all times the amount of pressure in the boiler; and a fusible plug to give warning in case of "low water."

Thus we see that in speaking of a boiler, not only the boiler proper is meant, but also the whole of its fixtures and belongings, of which the following is only a partial list:

Feed Pump,
Injector or Inspirator,
Check Valve,
Guage Cocks,
Glass Water Guage,
Try Cocks,
Blow-out Apparatus,
Blow-off Valve,
Safety Valve,
Scum Apparatus,
Steam Guage,
Fusible Plug,
Surface Blow Cocks,
Grate Bars,
Baffle or Shield Plates,
Mud Drum,
Feed Water Heaters,
Boiler Fronts,
Dead Plate,
Steam Pressure Recording Guage,
Drain Cock for Steam Guage,
Steam Trap,
Steam Whistle.

BOILER FIXTURES.

All these are attachments to the boiler proper, having direct reference to its internal functions; but in addition there are the lugs, pedestals, or brackets which support the boiler; the masonry in which it is set, with its binders, rods, and wall-plates; the boiler front, with its doors, anchor-bolts, etc.; the arch-plates, bearer-bars, grate-bars, and dampers, and last, but not least, the chimney. These are all equally necessary to enable the boiler to perform its duty properly. And besides, there are required fire-tools, flue brushes and scrapers, and scaling tools, with hose also, to wash out the boiler, to say nothing of hammers, chisels, wrenches, etc.

The fittings and attachments of the marine boiler are similar to those belonging to the land steam generators, and vary only in accommodating themselves to their peculiar surroundings.

The proper operation of the boiler as to efficiency and economy is largely dependent upon the number, appropriate proportion and harmony of action of its numerous attachments, and the utmost care and skill are requisite for designing and attaching them.

It must not be supposed that a complete list and description of all steam boiler attachments are here presented—that were a task beyond the limits of the entire volume.

BOILER FRONTS.

Boiler fronts are made in many different styles, almost every maker having some peculiar points in design that he uses on his own boilers and which nobody else uses.

In the illustrations here given may be seen the four principal designs:

1. The flush front as shown in Fig. 72.
2. The overhanging front as seen in Fig. 73.
5. The cutaway front, Fig. 74.
4. Fronts with breaching as shown in Fig. 75.

The flush front is one of the earliest forms of fronts, and though it often gives good satisfaction, yet it is liable to certain accidents.

BOILER FRONTS.

Front for Water Tube Boiler.—Fig. 71.

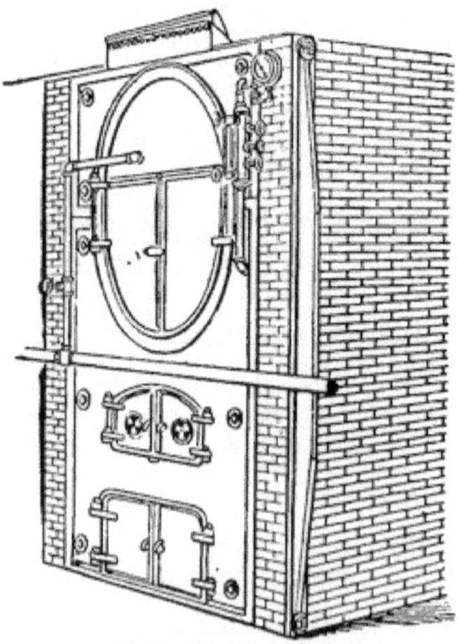

Flush Front.—Fig. 72.

As will be seen from cut 72, the front of the smoke arch, in this form of setting, is flush with the front of the brickwork, and the dry sheet just outside of the front head is built into the brickwork. The heat from the fire, striking through the brickwork, impinges on this sheet, which is unprotected by water on the inside. So long as the furnace walls are in proper condition the heat thus transmitted should not be sufficient to give trouble; but after running some time bricks are very apt to fall away from over the fire door, and thus expose portions of the dry sheet to the direct action of the fire, causing it to be burned or otherwise injured by the heat, and perhaps starting a leakage around the front row of rivets when the head is attached to the shell.

In the overhanging front this tendency is entirely prevented by setting the boiler *in such a manner that the dry sheet projects out into the boiler room.* If the brickwork over the fire door falls away when a boiler is set in this man-

BOILER FRONTS.

ner, the only effect is to slightly increase the heating surface. No damage can be done, since the sheet against which the heat would strike is protected by water on the inside.

The objection is sometimes raised against the projecting front, that it is in the way of the fireman. To meet this point and yet preserve all the advantages of this kind of front, the cutaway style has come into use. In this form the lower portion or the front sheet is cut obliquely away, so that at the lowest point the boiler projects but little beyond the brickwork.

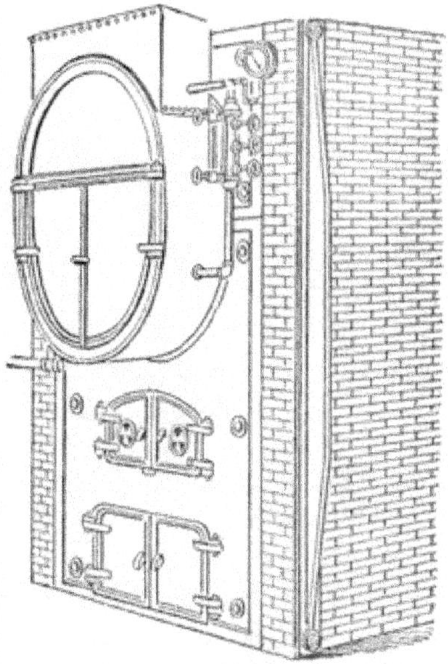

Overhanging Front.—Fig. 73.

It will be noticed that in the flush and overhanging fronts, the doors open sidewise, swing about on vertical hinges; in the cutaway front the best way to arrange the tube door is to run a hinge along the top of it, horizontally, and to have the door open upward. But with such a disposition of things the door is not easy to handle. For the purpose of support a hook and chain, hanging from the roof should be provided.

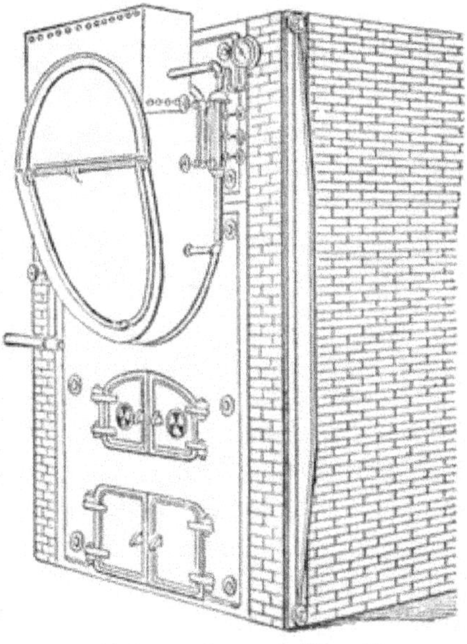

Cutaway Front.—Fig. 74.

BOILER FRONTS.

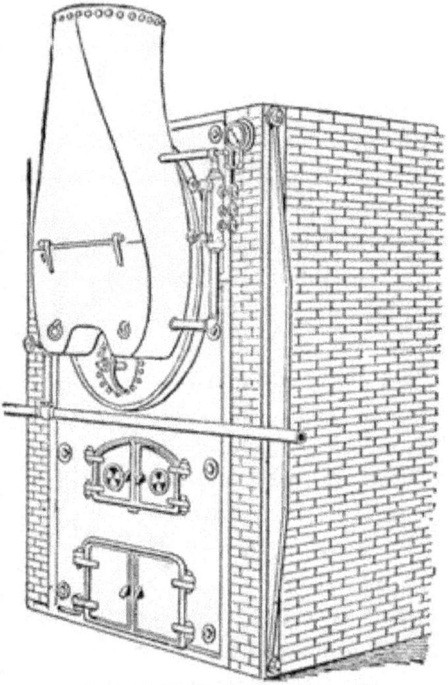

Front for Manhole.—Fig. 75.

Fig. 75 shows a boiler the setting of which is similar in general design to the other three, except that in the place of a cast-iron front it has bolted to it a sheet iron breeching that comes down over the tubes and receives the gases of combustion from them. In Fig. 75 a manhole is shown under the tubes. This, of course, is not an essential feature of the breeching, but it will be seen that manholes can readily be put below the tubes on fronts of this kind, in such a manner as to be very convenient of access.

In addition to these more general styles of boiler fronts, there are fronts designed particularly for patent boilers, water-front boilers, etc., which are made, very often, in ornamental and attractive designs. In Fig. 71 is shown a beautiful and appropriate design in use in connection with water tubular boilers.

FURNACE DOORS.

The chief points to be considered in the design of furnace doors are to prevent the radiation of heat through them, and to provide for the admission of air above the burning fuel in order to aid in the consumption of smoke and unburnt gasses.

In all cases where the doors are exposed to very rough usage —such, for instance, as in locomotive and marine boilers—the means for admitting air must be of the simplest, and consist generally of small perforations as shown in Fig. 76 which re-

FURNACE DOORS.

presents a front view, and section of the furnace door of a locomotive boiler. The heat from the burning fuel is prevented

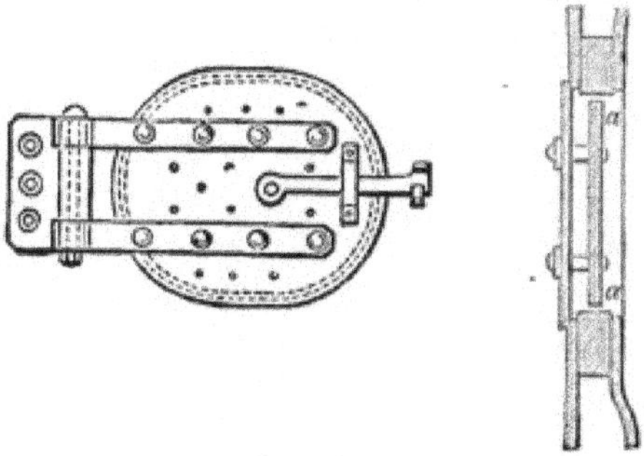

Fig. 76.

from radiating through the perforation in the outer door, by attaching to it a second or baffle plate, *a*, at a distance of about 1½ inches, the holes in which do not coincide in direction with the door proper. By the constant entry of cold air from the outside the greater part of any heat which may be communicated to the door by radiation or conduction is returned to the furnace.

Doors similar to the above provide for the constant addition of limited quantities of fresh air above the fuel, but in actual practice, however, air is only needed above the fire for a few minutes after fresh fuel has been thrown on the grates and then is required in considerable quantities. In the case of land boilers, the furnace doors of which undergo comparatively mild treatment, it is possible to introduce the necersary complications to effect this object.

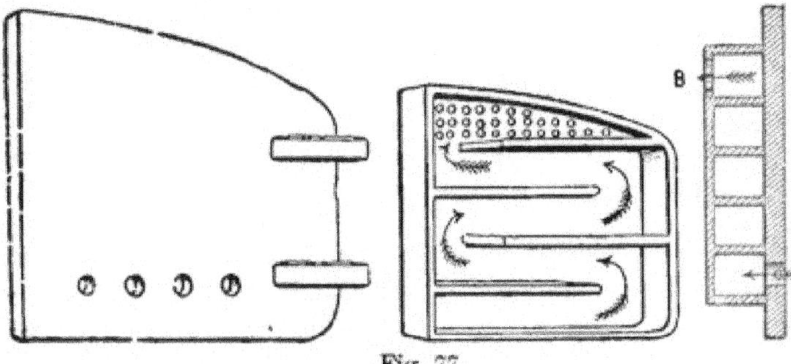

Fig. 77.

FURNACE DOORS.

Fig. 77 shows an arrangement largely in use in New England, in which, by means of a diaphram, the air is passed back and forth across the heated inner or baffle plate with the very best results.

The air is first drawn by the natural draught into the hollow space between the iron door and its lining, through a row of holes *A*, in the lower part of the door, controlled however, by a slide not shown in the cut, then caused to flow back and forth across the width of the door by simply arranged diaphrams, and finally injected into the furnace through a series of minute apertures drilled in the upper part of the door liner, as indicated in cut at *B*.

It will be seen that while the air may enter the door at a low temperature, it constantly becomes heated during its circulation until the instant it enters the furnace, it is ready to flash into flame with intense heat upon its incorporation with the expanding gasses of the furnace.

An arrangement in common use in Cornish and Lancashire boilers consists of a number of radial slits in the outer door which can be closed or opened at will in the same manner as an ordinary window ventilator. Other and more complicated arrangements have been frequently devised, which work admirably so long as they remain in order, but the frequent banging to which furnace doors are subjected, even in factory boilers, soon deranges delicate mechanism.

Furnace doors should be made as small as possible considering the proper distribution of fuel over the grate area, as otherwise the great rush of cold air, when the door is opened rapidly cools down the flues and does considerable injury to tube plates, etc.; for this reason it is desirable, when grates are over forty inches in width to have two doors to each furnace, which can be fired alternately.

The great loss arising from a rush of cold air on opening the furnace doors for replenishing the fires with fuel has led to costly experiments to produce "a mechanical stoker," or self boiler feeding arrangement for supplying the coal as needed.

FUSIBLE PLUGS.

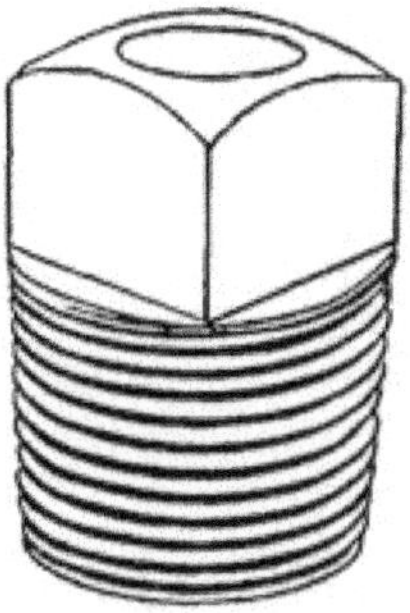

Fig. 78.

In some States the insertion of fusible plugs at the highest fire line in boilers is compelled by law under a heavy penalty. Its design is to give the most emphatic warning of low water, and at the same time relieve the boiler of dangerous pressure.

Figs. 78 and 79 exhibit two of the forms most commonly used, and on the succeeding page, in cut 80, is shown the device in operation where the water has sunk to a dangerously low level. In the illustration the device is shown in connection with a locomotive boiler, in the common tubular boiler the plug is usually inserted in the rear head of the boiler, so that in case of its operation it will not endanger the fireman.

These devices are designed to be screwed into the boiler shell at the safety line. The figs. 78 & 79 exhibits their construction. The part to be screwed into the boiler is called *the shell* and is commonly made of brass; the internal part is plug and is made of a soft metal like banca tin or a compound consisting of lead, tin and bismuth. This composition melts easily at the proper point to allow escape, where the water has sunk to a dangerously low level.

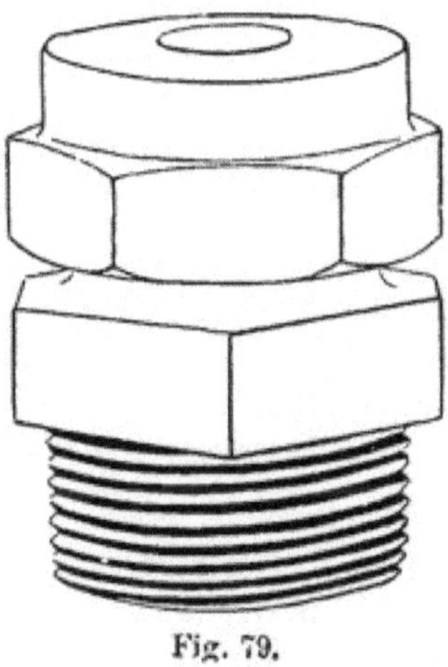

Fig. 79.

There is considerable diversity in the make up of the material used for filling the plug, which must not have its melting point at anything less than the temperature of the steam lest it should "go off" at the wrong time.

FUSIBLE SAFETY PLUG.

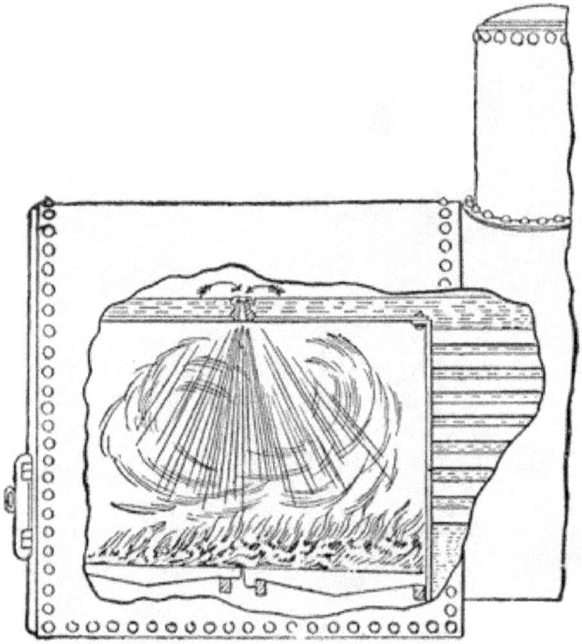

Fig. 80.

If the accident of low water occurs at a time where it is important to continue operations with the least possible delay, a pine plug may be driven in the opening left by the melting of the fusible metal. In any event it is but a short job to renew the fusible cap, it being only necessary to unscrew the nut and insert a new cap, the rest of the device remaining intact.

The plug should be renewed occasionally and the surface exposed inside the boiler be kept free from scale and deposit. It is to be understood that the fusible portion extends entirely through the shell of the boiler and when melted out makes a vent for the water or steam.

All marine boilers in service in the United States are required to have fusible plugs, one-half inch in diameter, made of pure tin, and nearly all first-class boiler makers put them in each boiler they build.

GRATE BARS.

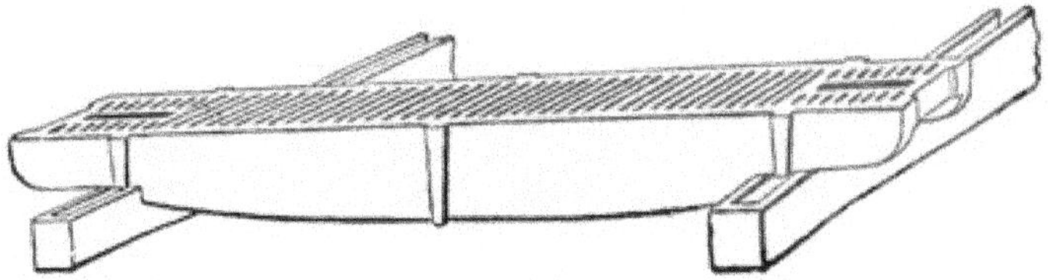

Fig. 81.

The Grate Bars are a very important part of the furnace appliances. These consist of a number of cast iron bars supported on iron bearers placed at and across the front and back of the furnace. Innumerable forms of grate bars have been contrived to meet the cases of special kinds of fuel. The type in common use is represented in Fig. 82.

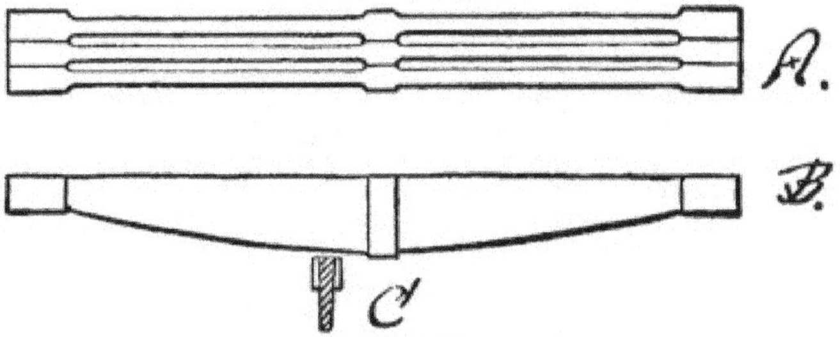

Fig. 82.

These cuts show a side view and a section of a single bar, and a plan of three bars in position. Each bar is in fact a small girder, the top surface of which is wider than the bottom. On each bar are cast lugs, the width of which determines the size of the opening for the passage of air. This opening varies in width according to the character of the fuel; for anthracite $\frac{3}{4}$ inch is a maximum, while for soft coals $\frac{5}{8}$ to $\frac{3}{4}$ inch is often used; for pea and nut coal still smaller openings than either of those are used, *i. e.*, $\frac{1}{4}$ and $\frac{3}{8}$ inches. For wood the opening should be a full inch in width.

For long furnaces the bars are usually made in two lengths, with a bearer in the middle of the grate, as shown in Fig. 83. As a rule long grates are set with a considerable slope towards

GRATE BARS.

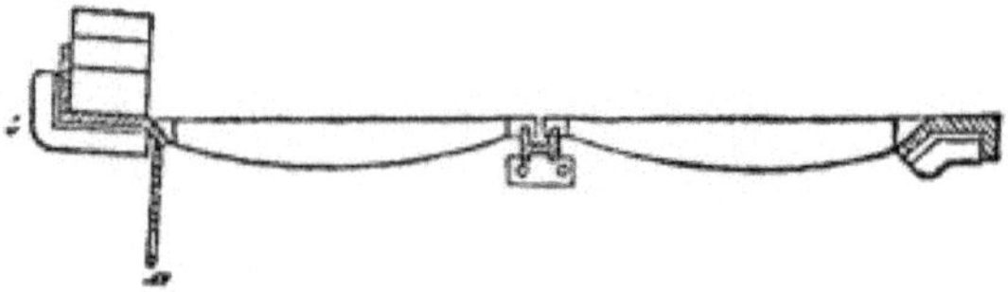

Fig. 83.

the bridge in order to facilitate the distribution of the fuel; an inch to a foot is the rule commonly approved.

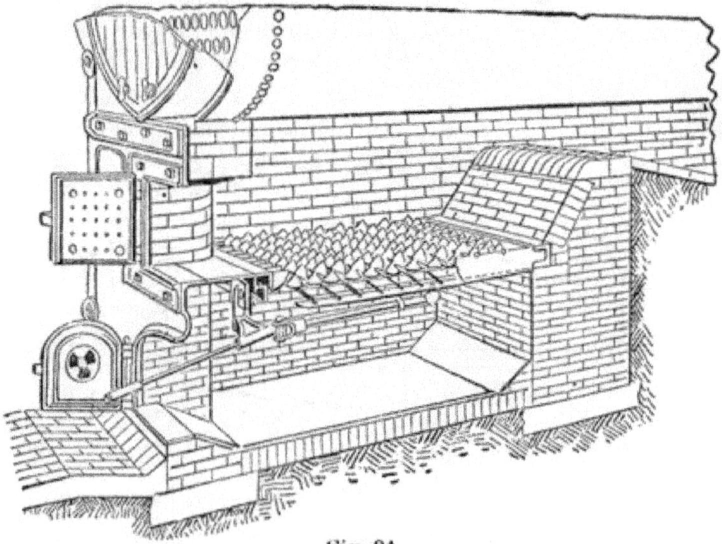

Fig. 84.

Rocking and shaking grates are now very extensively used; these combine a dumping arrangement, and very largely lessen the great labor of the fireman, and by allowing the use of slack and other cheap forms of fuel are very economical. Several patents are issued upon this form of grate bars all working on essentially the same principal. Fig. 84 exhibits an efficient form of a shaking grate. As shown in the cut, the grates are arranged to dump the ashes and clinkers. By the reverse motion the flat surface of the grates are restored.

Trouble with grate bars comes from warping or twisting caused by excessive heat, and burning out, produced by the same cause—this explains the peculiar shape in which grates are made—very narrow and very deep. A free introduction

GRATE BARS.

of air not only causes perfect combustion but tends towards the preservation of the bars.

Grate bars are usually placed so as to incline towards the rear, the inclination being from one to two inches; this facilitates somewhat the throwing of the coal into the furnace.

The proportion between grate and heating surface should be determined by the kind of fuel to be used. The greatest economy will be attained when the grate is of a size to cause the fire to be forced, and have the gases enter the chimney only a few degrees hotter than the water in the boiler.

If the grate is too large to admit of forcing the fire, the combustion is naturally slower, and consequently the temperature in the furnace is lower, and the loss from the escaping gases is greater.

It must be borne in mind that the only heat which can be utilized is that due to the difference in temperature between the fire and the water in the boiler. For example, if the temperature in the furnace be 975°, and the water in the boiler have a temperature due to 80 pounds of steam, viz.: 325°, it is evident that the heat which can be utilized is the difference between them, or $\frac{2}{3}$ of the total heat. Now if the fire be forced, and the furnace temperature raised to 2600°, $\frac{7}{8}$ of the total heat can be utilized; so it can be readily seen that the grate should be of such a size as to have the fire burn rapidly.

The actual ratio of grate to heating surface should not in any case be less than 1 to 40, and may with advantage, in many cases, be 1 to 50. This proportion will admit of very sharp fires, and still insure the greater portion of the heat being transmitted to the water in the boiler.

The water grate bars, invented in 1824, and since frequently applied to locomotives and marine boilers, do not seem to grow in popular favor, and are scarcely known in stationary boilers.

The objections urged against them are the expense of maintenance, their fittings and attachments, and the possibility of serious consequences should they rupture or burn out.

WATER GAUGE COCKS.

It is of the first importance that those in charge of a boiler shall know with certainty the position of the water level within the boiler.

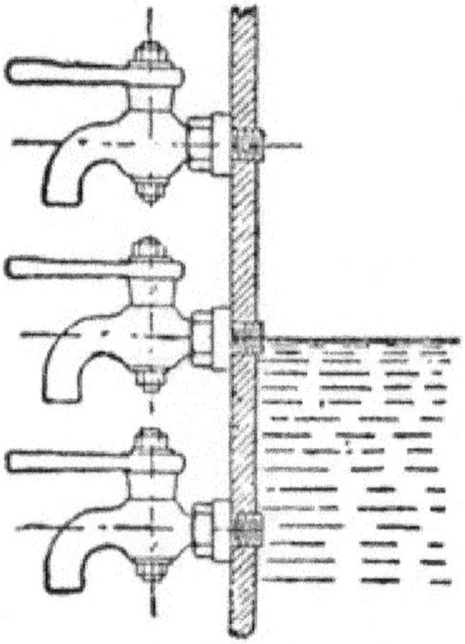

Fig. 85.

These attachments, also called Try cocks, are usually placed in a conspicuous and accessible position on the front of boilers. They are so arranged that one will blow only steam, one at the working level of the water, and the third at the lowest water level or say three inches above the highest point of the fire line of the boiler. The cut, Fig. 85, exhibits them as commonly arranged.

It is not essentially requisite, that the the cocks themselves should be placed at the point indicated, so long as they have pipes projecting internally into the boiler, with their ends corresponding to the height of water above mentioned. In order that these cocks may readily be cleaned out, a plug is usually fitted into bit of cock opposite the port or opening of the plug, upon removing which a pricker can be readily inserted.

The gauge or cocks should be tested many times each day, and when opened the top one should always give *steam* and the bottom one *water*. They should be allowed to remain open long enough to make sure whether steam or water is issuing from the cock. This is a matter of instruction, but the beginner with a little experience can detest the difference by the sound.

In so universal an appliance as this there are very many forms and arrangements, but they all work upon the same principle as stated above.

GLASS GAUGES.

These are the second and auxilliary arrangements for ascertaining the water line. Nearly all boilers are supplied with both try cocks and glass gauges, and so important is it considered to be correctly informed as to the water line that a third method consisting of a float which is carried on the water surface, is sometimes added to the two named.

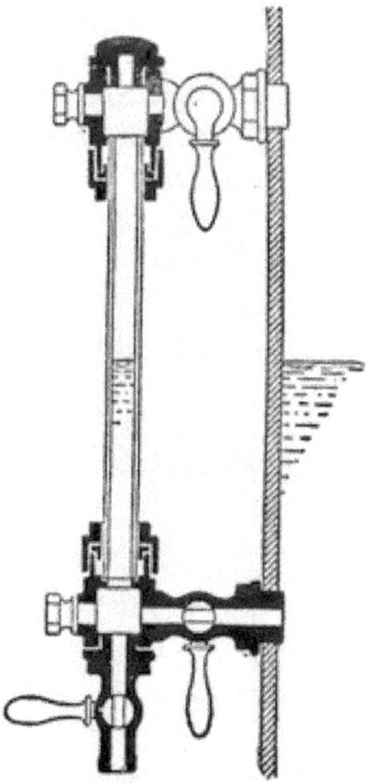

Fig. 86.

The glass water gauge *column* consists of an upright casting bolted to the front of the boiler, in which are fixed two cocks having stuffing boxes for receiving the gauge glass. The lower of these cocks is also fitted with a drain cock for blowing out the glass.

The try cocks are frequently placed on the above-mentioned standard or column.

The action of the gauge glass is to show the level of the water in the boiler by natural gravitation and the best position for it is in view of the engine room, as close to the boiler as possible and preferably in the middle line of its diameter, at such height that its lowest portion is about two inches above the highest part of the fire line of the boiler, and its center, nine inches above that, making the total visible portion of glass eighteen inches long.

Glass water gauges sometimes have pipe connections top and bottom. The object of this arrangement is to have an undisturbed water level in the glass by carrying one pipe to the steam dome and the other near to the bottom of the boiler; the one position not being so liable to be effected by foaming and the other by the boiling of the water. Cocks should always be fitted to the boiler ends of these pipes, in order that

WATER GAUGES.

in case of accident to the pipes, steam and water may be shut off.

The glasses are liable to burst and become choked up with dirt. The former defect is easily repaired by shutting off the cocks in connection with the boiler and putting in a new glass. The mud or sediment is cleaned out by opening the above-mentioned drain or blow out cock and allowing the steam or water, or both, to rush through the glass, which will effectually blow out all sediment and leave the glass in good condition again to show the height of the water in the boiler.

In opening the cocks connected with the glasses, it should be done cautiously, as the glass is liable to burst.

A strip of white running the whole length of the glass on the side toward the boiler is a great help in observing the variations of the water line in the tube.

It is not needed to remove the gauge glasses to clean them. There are good fixtures in the market that by taking out the plug in the top, the glass may be cleaned with a bit of wicking on the end of a stick. A slight scratch will break the glass, hence do not use wire. Use soft rubber gaskets when setting the glass, screw up until all leaking stops. Don't let the glass come in contact with the metal *anywhere.* Don't try to reset the glass with an old, hard gasket. Two glasses from the same bundle will not act alike.

The glasses used to show the water line are made of a soft glass known as "lead glass," and are easily cut, or broken square across. Most of them can be broken by filing a notch at the point at which it is necessary to break them. After filing the notch, place the thumbs as if you would break the glass; it will crack easily, and the fracture be straight and clean. If the tube be brittle, as some are, to avoid cutting the hands wrap two pieces of paper around the glass, each side of the notch. If the ends are rough or uneven, they can be made smooth by filing or by the grindstone.

The Manchester, Eng., Boiler Association attribute more accidents to inattention to water gauges than to all other causes put together. It is, therefore, of much importance that these glasses should be kept clean. It is not an uncommon thing to go into a boiler room and find that a leaky stuffing box has allowed the steam or water to blow out, and, by running down the outside of the glass, leave a deposit of lime scale. After this deposit has been formed, it is sometimes difficult to remove —and more than a few glasses have been broken by the engineer attempting to remove the scale. After this scale has once been formed, unless it is soft enough to be wiped off with a piece of waste, it is best to take the glass out and soak or wash it in a solution of one-half muriatic acid and one-half water until it is clean or the scale so softened that it may be readily wiped off. To prevent the scale from again forming and hardening, the glass should be dipped in glycerine before replacing.

THE MUD DRUM.

The mud drum is attached to a boiler with the expectation that it will catch and hold the larger portion of the sediment precipitated from the water. The mud drum to be effective should be protected from the heat of the fire, for so soon as it receives sufficient heat to boil the water within it can no longer serve the purpose for which it was intended as all the sediment which may have gathered would be expelled by the ebullition of the water. When the drum is located under the boiler it is not in a good position to catch the sediment, as the boiling water produces sufficient current to carry the sediment to the top, or keep it violently agitated, so that there is little opportunity for it to be deposited any where so long as the boiler is making steam. Afterward when the water is quiet the sediment for the most past is deposited on the tubes and the curve of the shell; the small portion falling into the neck of the drum serves principally to show the inefficiency of the device. Located under the boiler as it generally is, makes it extremely difficult to get at for examination, and as a consequence of its being enclosed, as it must be, to be of much importance, it is subject to greater deterioration than would otherwise be the case, and as the enclosure to be most efficient would enclose the neck also, the

THE MUD DRUM.

difference of expansion at or near the junction would soon produce leaking if not worse. When the mud drum is located outside the boiler walls where it would be most efficient, if properly connected, it loses its identity and becomes a mechanical boiler cleaner. In consequence of these drawbacks the mud drum is becoming antiquated as a boiler appliance, and is now seldom used.

BAFFLE PLATES.

These are a device sometimes used inside steam boilers to check the too sudden flow of steam towards the exit pipe, they are simply plate to baffle the rush of the steam so as to avoid foaming.

In Fig. 90 baffle plate is illustrated by the division casting against which the steam strikes on its passage from the boiler to the engine. The liners or inner plates of the boiler doors are baffle plates.

DEAD PLATE.

This is a flat plate of iron immediately inside the furnace door and is used in many boilers in order to insure the more perfect combustion of the coal.

When the fresh fuel is laid on, it is placed on the dead plate instead of on the grate; in this position the coal is coked, the gasses from the coal being ignited as they pass over the already intensly hot fuel in the furnace, the fuel from the dead plate is pushed forward to make place for another charge to be put on the dead plate. But more frequently, as elswhere described, the fuel is thrown over and across the dead plate directly upon the hot fire.

STEAM WHISTLES.

These are of two kinds, known as the bell-whistle and organ-tube whistle; the latter is now fast superceding the former on account of the simplicity of construction and superior tone. An improved form has a division in the tube so as to emit two distinct notes, which may be in harmony, or discord, and when sounded together may be heard a long distsnce.

It is important that the whistle shall sound as soon as the steam is turned on; to ensure this great care must be taken to keep the whistle-pipe free from water.

THE STEAM GAUGE.

The principle of construction of the dial steam gauge is, that the pressure may be indicated by means of a pointer in a divided dial similar to a clock face, but marked in division, indicating pounds pressure per square inch instead of hours and minutes.

Figs. 87 and 88 show the ordinary style of gauge which consists of an elliptical tube, connected at one end to a steam pipe in communication with the boiler pressure and at the other end with gearing to a pointer spindle as shown in cut.

An inverted syphon pipe is usually formed under the gauge, its object being to contain water and thus prevent the heat of the steam injuring the machinery of the gauge, or distorting its action by expansion.

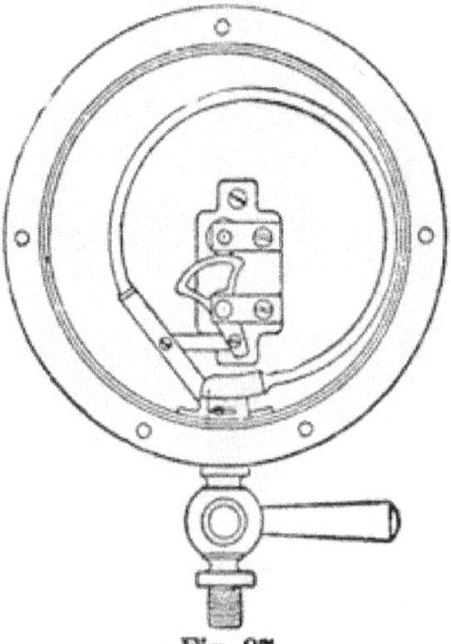
Fig. 87.

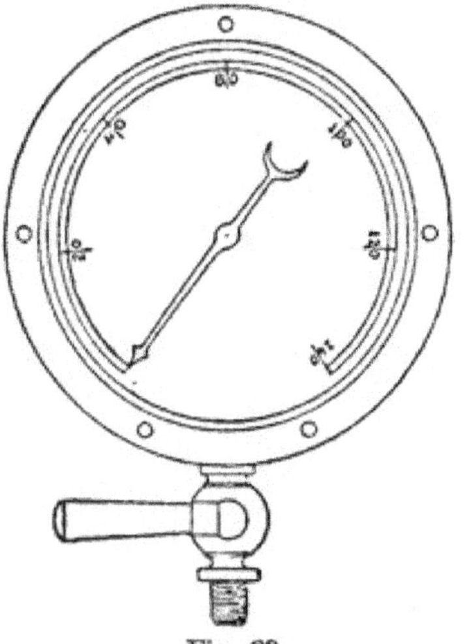
Fig. 88.

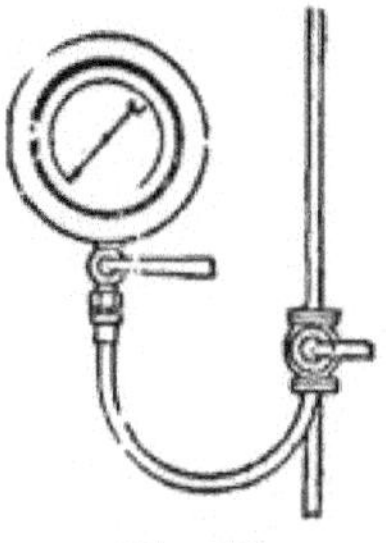
Fig. 89.

A small drain cock should be fitted to the leg of the syphon of a steam gauge, leading to the boiler, at a level with the highest point the water can rise in the other leg, otherwise an increased pressure will be indicated, due to the head of water which would otherwise collect in the boiler leg of the syphon.

Steam gauges indicate the pressure of steam

above the atmosphere only, the total pressure being measured from a perfect vacuum which will add 14$\frac{7}{10}$ lbs. on the average to the pressure shown on the steam gauge.

These gauges are apt to get out of order in consequence of water lodging in the end of the heat tube and corroding the latter. It may be easily known when they are out of order by raising the pressure of the steam in the boiler and watching when it commences to blow off at the safety valve, and then noting the position of the index finger. The pressure registered by the finger should, of course, then correspond with the known blow off pressure of the valves; if it does not, one or the other or both of these instruments must be out of order; therefore, when this is the case and a disagreement occurs, the steam gauge may be presumed to need correction.

It should also be noted that the steam gauge finger points to zero when steam pressure is cut off. A two-way cock should be used for closing the connection between the steam gauge and the boiler, and at the same time to let air into the steam gauge.

The steam should never be allowed to act directly on a steam gauge when located in cold situations where they are liable to freeze. The valve on the boiler should be closed and the water allowed to drip out, and, before the steam is turned on from the boiler, the drip on the gauge should be closed, in order that sufficient steam may be condensed in the pipe to furnish the quantity of water necessary to keep the steam from striking the gauge.

A ready method for being always able to prove the correctness of your steam gauge.

When steam is at some point not over half the usual pressure, place the ball on the safety valve at the point where it commences to blow off and mark the place. Move the ball twice as far from the fulcrum as this mark, and it should blow off at twice the pressure as indicated by the gauge, or it is not right. Any other relative distance may be used to advantage.

STEAM SEPARATOR.

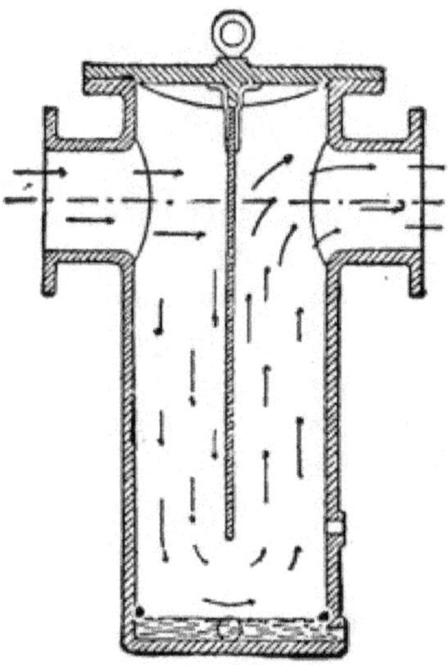

Fig. 90.

This appliance, which is also called an interceptor or catch water, is generally a T shaped pipe.

This, although not a boiler fixture or fitting, is intimately connected with them: it is an appliance fast coming into use both for land and marine engines, to guard against the danger to steam engine cylinders arising from "the priming" of the boilers when the steam is used at a high pressure with high speed of the piston.

The separator is usually placed in the engine room, so as to be well in sight. The steam is led down the pipe round a diaphragm plate and then up again to the engine steam pipe. By this means any priming or particles of water that may be brought from the boiler with the steam will fall to the bottom of the interceptor or catch water, from whence it can be blown out, according to the arrangement of the pipes, by opening the drain cock fixed on the bottom. It has a water gauge fixed on the lower end, so as to show whether water is accumulating; and the engineer's attention is required to see that this water is from time to time blown off.

In the illustration, Fig. 90, is shown the simplest form in which the device can be made. The arrows exhibit the direction in which the steam travels, the aperture whence the water is to be blown out and the place for attachment of a water column. In practical construction the separator should have a diameter twice that of the steam pipe and be 2½ to 3 diameters long. It is often made with a round top and flat bottom and sometimes with both ends hemispherical. The division plate should extend half the diameter of the steam pipe below the level of the bottom of the steam pipe.

THE SEPARATOR.

In Fig. 91 is shown an improved form of a steam separator which consists of a shell or casing in which there is firmly secured a double-ended cone. On this cone there are cast a number of wings, extending spirally along its exterior. On entering the separator the steam is spread and thrown outward by the cone and given a centrifugal motion by the spiral wings. These wings are constructed with a curved surface.

It will be noticed that the steam on entering the separator is immediately expanded from a solid body into an annular space of equal volume to the steam pipe, whereby its particles are removed from the center and thus receive a greater amount of centrifugal motion. The entrained water or grease, etc., is thus precipitated against, and flows along the shell of the separator, and is collected in a well of ample proportions at base of separator, where it is entirely isolated from the flow of dry steam.

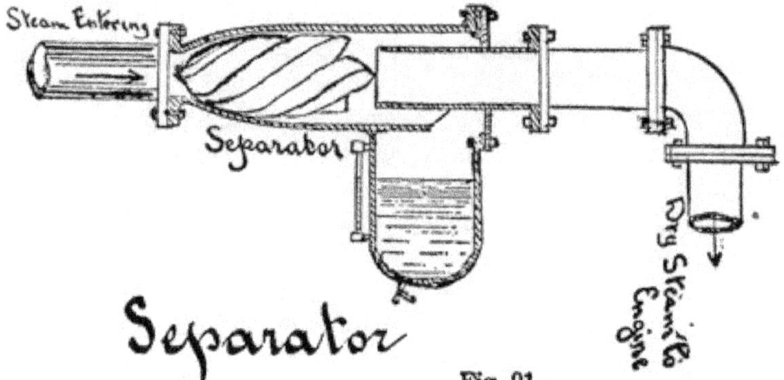

Fig. 91.

SENTINEL VALVE.

It was formerly required for each marine boiler to have a small valve loaded with a weight to a few pounds per square inch above the working pressure, so that in case of the safety valves sticking fast and the gauge being false, an alarm might be given when there was an excess of pressure. Such valves were about ¾ inch in diameter and sometimes as small as ⅜. An arrangement of a small safety valve attached to a whistle has been introduced, but with advances in other directions relating to safety these specialties are now getting to be only known by name.

DAMPER REGULATORS.

These are well-known devices for so controlling the draught of the chimney that the steam pressure in the boiler will be increased or decreased automatically, that is, without the aid of a person. The regulator shown in Fig. 92, which is one of many excellent forms on the market, has the power to move the damper in both directions by water pressure, exerting a force on the end of the lever of nearly 200 lbs., thus compelling a certain and positive motion of the damper when a variation in the boiler pressure takes place. It will open or close the damper upon the variation of less than one pound of pressure. The close regulation affords a test for the correctness of the steam gauge.

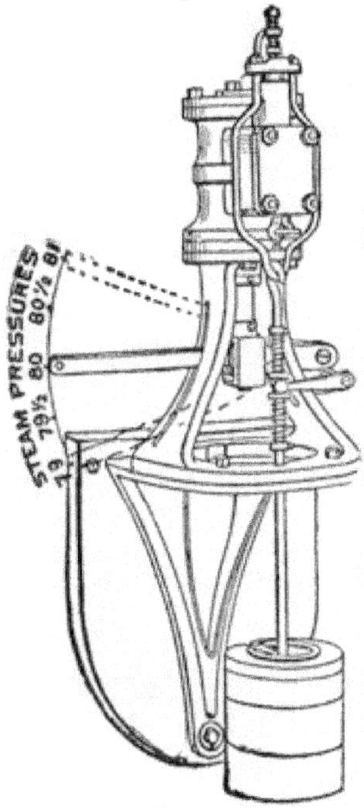

Fig. 92.

This regulator, by using the water pressure from the boiler as a motive power, becomes a complete engine without the connecting rod and crank, having a balanced piston valve, the valve stem of which is enlarged where it passes through the upper end of the chest into a piston of small area, working in a small open ended cylinder cast on the chest. The pressure forcing this piston outward is counterbalanced by weights as shown in illustration.

The differential motion is accomplished by the device shown at the top of small cylinder.

FUEL ECONOMIZER AND FEED WATER PURIFIER.

This device, shown in Fig. 93, is designed to utilize the waste products of combustion as they pass from the furnace to the chimney. Its use permits a high and consequently efficient temperature under the boilers and yet saves the excess of heat. It acts also as a mechanical boiler cleaner, furnishing a settling

FUEL ECONOMIZER AND FEED WATER PURIFIER.

chamber for the deposit of the impurities separated by the heat which nearly equals that of the live steam in the boiler. This device adds largely to the water capacity of the boiler, frequently containing one-half the weight of the water held in the boiler itself.

It will be readily understood that the openings between the vertical tubes are ample for the chimney flue area and that the device is located between the chimney and the boiler, with the waste furnace heat passing between the tubes.

Fig. 93.

The economizer shown in Fig. 93 consists of sections of vertical $4\frac{1}{2}''$ boiler tubes fitted to their top and bottom headers by taper joints. The top headers are provided with caps over each tube to permit cleaning out the sediment and remove and replace any tube that may become damaged. The several top headers are connected together at one end by lateral openings and the bottom headers are also connected as shown in cut, having hand holes opposite each bottom header to provide for cleaning out.

FUEL ECONOMIZER AND FEED WATER PURIFIER.

Mechanical scrapers are made to travel up and down each tube to keep them clear of soot. These are controlled by an automatic mechanism and driving head, as shown in the Fig. 93.

The important features about the economizer are its adaptability to any type of boiler, the great saving attained by utilizing that heat which has necessarily been an almost total waste, the purifying of the water by means of the intense heat and slow circulation obtained.

SAFETY VALVES.

The most important fitting upon a steam boiler is the safety valve.

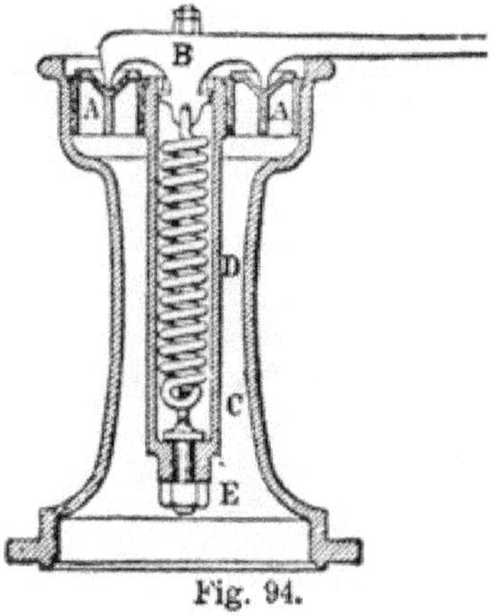

Fig. 94.

It may also be defined, as applying to all valves, that the *seat of the valve* is the fixed surface on which it rests or against which it presses, and *the face of a valve* is that part of the surface which comes in contact with the seat. The *spindle* is the small rod, sometimes made of composition bronze, which projects upwards or downward from the middle of the valve, and so arranged that it causes the valve to raise and drop evenly upon its seat. The seat is preferably leveled at an angle of forty-five degrees.

Generally speaking, the safety valve is a circular valve seated on the outside of the boiler, and weighted to such an extent, that when the pressure of the steam exceeds a certain point, the valve is lifted from its seating and allows the steam to escape. Safety valves can be loaded directly with weight valves, or the load can be transmitted to the valve by a lever. Again, the end of the lever is sometimes held down by a simple weight attached to it, a plan generally adopted in land

THE SAFETY VALVE.

boilers; while sometimes as in the case of locomotive and marine boilers, the lever is weighted by means of a spring, the tension of which can be adjusted.

A valve much used in locomotives is shown in Fig. 94.

It consists really of two valves *A*, *A*, placed side by side, at a little distance apart. A cross piece *B* bears upon each and to the cross piece is attached a powerful spiral spring *D*, the lower end of which is so fixed at *C* that its tension can be adjusted by means of a set screw at *E* which is out of reach of the engine driver. Before the valves can rise they have to overcome the resistance of the spring to which the pressure is communicated by the cross piece. The spring is attached to the cross piece below the bearing points of the cross piece on the valves, hence if one of the valves should rise from its seating before the other, the spring leans a little towards this latter, easing the pressure on it, and allowing it to open. The rise of the valve from its seating is much greater with these directly loaded valves than when the pressure is transmitted through a lever, and thus the steam steam escapes with much greater rapidity.

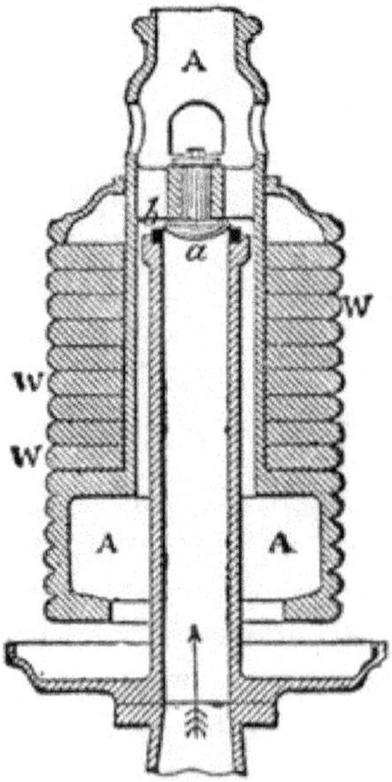

Fig. 95.

A "pop" safety valve is a common form of safety valve and takes its name from the fact that it takes a little more pressure to raise it off its seat than what it is set at, consequently it releases itself with a "pop."

Fig. 95 shows a form of dead weight safety valves when *a* is the valve which rests on the seating *b*.

The valve is attached to the circular casting *A*, *A*, *A*, so that both rise and fall together. The weights W. W., etc., are disposed on the casting in rings, which can be adjusted to the desired blow off pressure. Owing to the center of gravity of the casting and weight being below the valve, the latter requires no guides to keep it in position. This is a great advantage as guides frequently stick, and prevent the valve from acting. Another advantage of this form of valve is, that it is difficult to tamper with. For instance, a four inch valve, intended to blow off at 100 lbs. per square inch would require weight of over 1,200 lbs., which require a considerable bulk. An unauthorized addition of a few pounds to such a mass would make no appreciable addition to the blowing off pressure, while any effectual amount added to the weight would be immediately noticed. It is quite different with the lever safety valve about to be described, a small addition to the weight at the end of the lever is multiplied several times at the valve.

U. S. Rules Relating to Safety Valves.

Extract from rules and regulations passed and approved Feb. 25, 1885, by the United States Board of Supervising Inspectors of Steam Vessels:

Section 24. "Lever safety valves to be attached to marine boilers shall have an area of not less than one square inch to two square feet of the grate surface in the boiler, and the seats of all such safety valves shall have an angle of inclination of forty-five degrees to the centre line of their axis.

"The valves shall be so arranged that each boiler shall have one separate safety valve, unless the arrangement is such as to preclude the possibility of shutting off the communication of any boiler with the safety valve or valves employed. This arrangement shall also apply to lock-up safety valves when they are employed.

"Any spring-loaded safety valves constructed so as to give an increased lift by the operation of steam, after being raised from their seats, or any spring-loaded safety valve constructed in any other manner, or so as to give an effective area equal to that of the aforementioned spring-loaded safety valve, may be used in lieu of the common lever-weighted valve on all boilers on steam vessels, and all such spring-loaded safety valves shall

U. S. RULES RELATING TO SAFETY VALVES.

be required to have an area of not less than one square inch to three square feet of grate surface of the boiler, and each spring-loaded valve shall be supplied with a lever that will raise the valve from its seat a distance of not less than that equal to one-eighth the diameter of the valve-opening, and the seats of all such safety valves shall have an angle of inclination to the centre-line of their axis of forty-five degrees. But in no case shall any spring-loaded safety valve be used in lieu of the lever-weighted safety valve, without first having been approved by the Board of Supervising Inspectors."

The following size "Pop" Safety Valves are required for boilers having grate surfaces as below:

2	inch	"Pop"	Valve	for	9.42	square feet of	grate surface.
2½	"	"	"	"	14.72	"	"
3	"	"	"	"	21.20	"	"
4	"	"	"	"	37.69	"	"
5	"	"	"	"	58.90	"	"
6	"	"	"	"	84.82	"	"

PROFESSOR RANKIN'S RULE.—Multiply the number of pounds of water evaporated per hour by .006, and the product will be the area in square inches of the valve.

The U. S. Steamboat Inspection Law requires for the common lever valve one square inch of area of valve for every two square feet of area of grate surface.

United States Navy Department deduced from a series of experiments the following rule: Multiply the number of pounds of water evaporated per hour by .005, and the product will be the area of the valve in square inches.

Rule adopted by the Philadelphia Department of Steam Engine and Boiler Inspection:

1. Multiply the area of grate in square feet by the number 22.5. 2. Add the number 8.62 to the pressure allowed per square inch. Divide (1) by (2) and the quotient will be the area of the valve in square inches. This is the same as the French rule.

SAFETY VALVE RULES.

The maximum desirable diameter for safety valves is four inches, for beyond this the area and cost increase much more rapidly than the effective discharging around the circumference.

There should not be any stop valve between the boiler and safety valve.

The common form of safety valve is shown in Fig. 96.

Here the load is attached to the end *B* of the lever *A*, *B*, the fulcrum of which is at *c*. The effective pressure on the valve, and consequently the blowing off pressure in the boiler, can be regulated within certain limits, by sliding the weight *W* along the arm of the lever. In locomotive engines, as well as on marine boilers, the weight would on account of the oscillations, be inadmissible and *a spring* is used to hold down the lever. The pressure on the valve can be regulated by altering the tension of the spring.

In the calculations regarding the lever safety valve, there are five points to be determined, and it is necessary to know four of these in order to find the fifth. These are: (1) The Steam Pressure, (2) The Weight of Ball, (3) The Area of Valve, (4) The Length of Lever, (5) The Fulcrum.

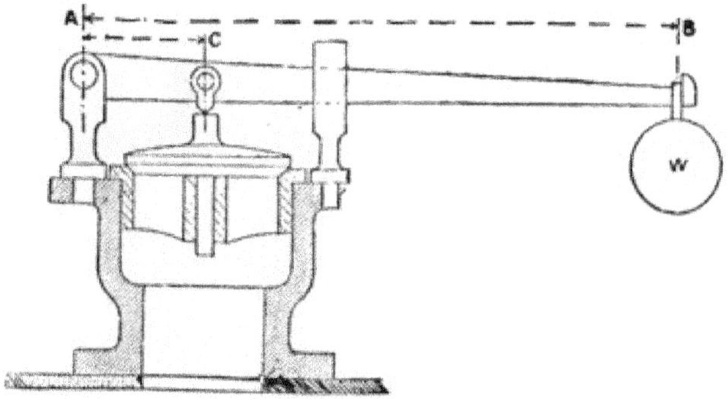

Fig. 96.

In making these calculations it is necessary to take into account the load on the valve due to the weight of the valve-stem and lever. The leverage with which this weight acts is measured by the distance of its centre of gravity from the fulcrum. The centre of gravity is found by balancing the lever on a knife edge, and the weight of the valve-stem and

SAFETY VALVE RULES.

lever can be found by actual weighing. This load can also be found by attaching a spring balance to the lever exactly over the centre of the valve stem when they are in position. The following examples will be computed under these conditions: (1) Steam Pressure, 120 pounds; (2) Weight of Ball, 100 pounds; (3) Weight of Valve and Lever, 60 pounds, weighed in position; (4) Length of Lever, 45 inches; (5) Length of Fulcrum, 5 inches; (6) Area of Valve, 8 square inches.

To find the area of the valve:

RULE.—Multiply the length of the lever by the weight of the ball, and divide the product by the fulcrum, and to the quotient add the weight of the valve and lever, and divide the sum by the steam pressure.

Example.

```
                              45   inches, length of the lever,
                             100 pounds, weight of the ball.
          Fulcrum, 5 in.)4500
                             900
                              60 pounds, weight of valve and lever,
Steam pressure 120 lbs.)960(8 square inches, area of valve.
                             960
```

To find the pressure at which the valve will blow off:

RULE.—Multiply the length of the lever by the weight of the ball; divide this product by the fulcrum, and to the quotient add the weight of the lever and valve, and divide the sum by the area of the valve.

Example.

```
                      45   inches, length of lever,
                     100 pounds, weight of ball,
Fulcrum 5 in.) 4500
                     900
                      60 pounds, weight of valve and lever,
Area of Valve 8) 960
                     120 pounds, pressure at which valve will blow.
```

SAFETY VALVE RULES.

To find the weight of ball:

RULE.—Multiply the steam pressure by the area of the valve, and from the product substract the weight of the valve and lever, then multiply the remainder by the fulcrum, and divide the product by the length of the lever.

Example.

```
                              120 pounds, steam pressure,
                                8 inches, area of valve,
                              ---
                              960
                               60 pounds, weight of valve and lever.
                              ---
                              900
                                5 inches, fulcrum,
                              ----
Length of lever, 45 in.)4500
                              ----
                              100 pounds, weight of ball.
```

To find the length of lever:

RULE.—Multiply the steam pressure by the area of the valve, and from the product substract the weight of the valve and lever, then multiply the remainder by the fulcrum, and divide the product by the weight of the ball.

Example.

```
      120 pounds, steam pressure,
        8 inches, area of valve,
      ---
      960
       60 pounds, weight of valve and lever.
      ---
      900
        5
      ----
100)4500(45 length of lever.
```

Every boiler should be provided with two safety valves, one of which should be put beyond the control of the attendant. The size of the openings depends of course, upon the steam producing power of the boiler.

POINTS RELATING TO SAFETY VALVES.

Safety valves that stick will do so even though tried every day, if they are simply lifted and dropped to the old place on the seat again. *If a boiler should be found with an excessively high pressure, it would be one of the worst things to do to start the safety valve from its seat unless extra weight was added,* for should the valve once start, it would so suddenly relieve the boiler of such a volume of steam as would cause a rush of water to the opening, and by a blow, just the same as in water hammer, rupture the boiler.

Such a condition is very possible to occur of itself when a safety valve sticks. The valve holds the pressure, that gets higher and higher, until so high that the safety valve does give way and allows so much steam to escape that the sudden changing of conditions sets the water in motion, and an explosion may result.

The noise made by a safety valve when it is blowing off may be regarded in two ways. First, by it is known that the valve is capable of performing its proper function, and that there is, therefore, a reasonable assurance that no explosion will result from excessive pressure of steam or other gas, and on the other hand too much noise of this kind indicates wasted fuel.

The hole of the safety valve may be 2, 3 or 4 inches; that does not say that the area is 3.1416, 7.06 or 12.56 square inches, but the area is that which is inside of the joint. The valve opening may be, say 2 inches, but *the circle of contact of valve to seat* may be of an average diameter of 2⅛ inches, if so, all the close calculations otherwise will not avail. In the first place, the area of 2 inches equals 3.1416; that of 2⅛ diameter equals 3.5466, showing a difference of .4 square inches.

NOTE.

Very extended rules issued by the U. S. Government for calculating the safe working pressure, dimensions and proportions of the safety valves for marine boilers are reprinted in "Hawkins' Calculations" for engineers.

POINTS RELATING TO SAFETY VALVES.

When a safety valve is described as a "2 inch safety valve," etc., it means that 2 inches is *the diameter* of the pipe; hence the following rule and examples for finding the area.

Rule for finding Area of Valve Opening.

Square the diameter of the opening and multiply the product by the decimal .7854.

Example.

What is the area of a 3 inch valve? Now then:

$3 \times 3 = 9 \times .7854 = 7.\frac{06}{100}$ square inches, Ans.

Note.—A shorter method of calculationg by .7854 in larger sums is to multiply by 11 and divide by 14, for decimal .7855 = the fraction $\frac{11}{14}$th. Note: .7854 is the area of a circular inch.

When valves rise from their seats under increasing steam pressure they do so by a constantly diminished ratio which has been carefully determined by experiment and reduced to the following table.

Pressure in Lbs.	Rise of Valve.	Pressure in lbs.	Rise of Valve.
12	1-36	60	1-86
20	1-48	70	1-132
35	1-54	80	1-168
45	1-65	90	1-168
50	1-86		

The following useful table was prepared by the Novelty Iron Works, New York.

Boiler Pressure in Lbs. Above the Atmosphere.	Area of Orifice in Sq. In. for Each Sq. Ft. of Heating Surface.	Boiler Pressure in Lbs. Above the Atmosphere.	Area of Orifice in Sq. In. for Each Sq. Ft. of Heating Surface.
0.25	.022794	40.	.001723
0.5	.021164	50.	.001389
1.	.018515	60.	.001176
2.	.014814	70.	.001015
3.	.012345	80.	.000892
4.	.010582	90.	.000796
5.	.009259	100.	.000719
10.	.005698	150.	.000481
20.	.003221	200.	.000364
30.	.002244		

FEED WATER HEATERS.

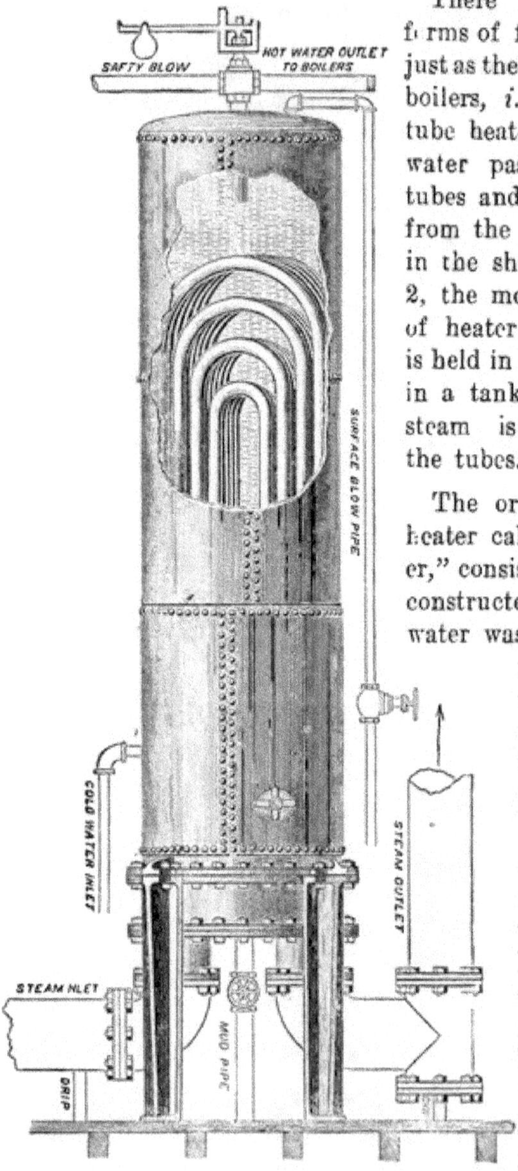

Fig. 97.

There are two general forms of feed water heaters just as there are two of steam boilers, *i. e.*, 1, the water tube heater, where the feed water passes through the tubes and extracts the heat from the exhaust contained in the shell of the heater. 2, the more approved form of heater where the water is held in the shell as it were in a tank, and the exhaust steam is passed through the tubes.

The original feed water heater called a "pot heater," consisted of a vessel so constructed that the feed water was sprayed through the exhaust steam into a globe formed tank, from the bottom of which the heated water was pumped into the boiler; its name was originally the "pot heater," but as it was open to the air through the exhaust pipe, it was, with its successively improved forms called the open heater.

FEED WATER HEATERS.

All the heat imparted to the feed water, before it enters the boiler, is so much saved, not only in the cost of fuel, but by the increased capacity of the boiler, as the fuel in the furnace will not have this duty to perform. There are two sources of waste heat which can be utilized for this purpose: the chimney gases and the exhaust steam. The gases escaping to the chimney after being reduced to the lowest possible temperature contain a considerable quantity of heat. This waste of heat energy may be largely saved by the device illustrated on page 186.

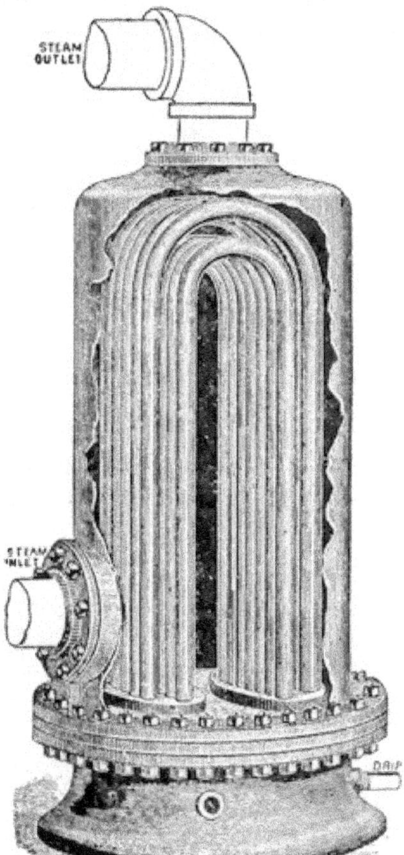

Fig. 98.

How much saving is obtained under any given condition is a question requiring for its solution a careful calculation of all of the conditions which have a bearing on the subject. Exhaust steam under atmospheric pressure only has a sensible temperature of 212 degrees, but exhaust steam contains also a large number of heat units which are given up when the steam is condensed into water; for this reason it might be thought possible to raise the temperature of the feed water a few degrees higher even than the sensible temperature of the exhaust steam. But this should not be expected, on account of the radiation of heat that would occur above that of the steam.

The steam which escapes from the exhaust pipe dissipates into the atmosphere or discharges into the condenser over nine-

FEED WATER HEATERS.

tenths of the heat it contained when leaving the boiler. This can be best utilized by *exhaust feed water heaters*, for the use of live steam heaters represents no saving in fuel, as all the heat imparted to the feed water by their use comes directly from the boiler. The purpose for which they are used is to elevate the temperature of the feed water above the boiling point, so as to precipitate the sulphate of lime and other scale forming substances, and prevent them from entering the boiler. Neither does the heat in the feed water introduced by an injector represent saving, as it comes from the boiler and was generated by the fuel.

It is important to note these two statements: 1, That neither live steam feed water heaters, nor 2, injectors save the heat from the escaping steam.

It is also well to remember that it requires *a pound of water* to absorb 1.146 heat units, and that this quantity of heat is distributed through the whole quantity of water, and *as a pound of steam is the same as a pound of water*, it may be understood that at 212° each pound of exhaust steam contains 1,146 heat units; ten pounds of steam contain 11,460 heat units distributed through the mass, etc.: thus, to explain still further:

To evaporate water into steam, it must first be heated to the boiling point, and then sufficient heat still further added to change it from the liquid to the gaseous state, or steam. Take one pound of water at 32 degrees and heat it to the boiling point, it will have received 212° — 32° = 180 heat units. A heat unit being the amount of heat necessary to raise one pound of water through one degree at its greatest density. To convert it into steam after it has been raised to the boiling point, requires the addition of 966 heat units, which are called latent, as they cannot be detected by the thermometer. This makes 180+966=1146 heat units, which is the total heat contained *in one pound of water* made into steam at the atmospheric pressure. And at atmospheric density the volume of this steam is equal to 26.36 cubic feet, and this amount of steam contains 1,146 units of heat, distributed throughout the whole quantity, while the temperature at any given point at

FEED WATER HEATERS.

which the thermometer may be inserted is 212 degrees. If two pounds of water be evaporated, making a volume of 52.72 cubic feet, then the number of heat units present would be doubled, while the temperature would still remain at 212, the same as with one pound.

If by utilizing the heat that would otherwise go to waste, the temperature of the feed water is raised 125 degrees, the saving would be $\frac{125}{1146}$ of the total amount of heat required for its evaporation, or about 11 per cent. Thus it can be seen the percentage of saving depends upon the initial temperature of the feed water, and the pressure at which it is evaporated.

For example, a boiler carrying steam at 100 pounds pressure has the temperature of the feed water raised from 60 to 200 degrees, what is the percentage of gain ?

By referring to a table pressure of "saturated steam," it will be seen that the total heat in steam at 100 pounds pressure is 1185 heat units. These calculations are from 32 degrees above zero, consequently the feed must be computed likewise.

In the first case, the heat to be supplied by the furnace is the total heat, less that which the feed water contains, or 1185—28=1157 heat units. In the second case it is 1185—168=1017 heat units, the difference being 1157—1017=140, which represents a saving of $\frac{140}{1157}$, or about 12 per cent.

Where feed water is heated no more than 20 degrees above its normal temperature the gain effected cannot amount to more than 2%, not sufficient to pay for the introduction and maintenance of a feed water heating device, no matter how simple, but if the temperature of the water can be increased 60 degrees the gain will be in the neighborhood of 5%. To make feed water heating practical and economical it would be necessary to increase the temperature of the water about 180 degrees at least, and to do this, using the exhaust from a non-condensing engine without back pressure, would require such a capacity of heater as would give fully 10 square feet of heating surface to each horse power of work developed, and to raise the temperature above this would require a certain amount of back pressure or an increased capacity of heater, so that the subject

FEED WATER HEATERS.

resolves itself into a question of large capacity of heater, or a higher temperature of the exhaust steam, which could only be obtained through a given amount of back pressure.

In the same way has been calculated the following table, showing percentages of saving of fuel by heating feed-water to various temperatures by exhaust steam, otherwise waste:

Percentage of saving. (*Steam at 60 pounds gauge pressure.*)

Final Temp. Fahr.	Initial Temperature of Water (Fahrenheit).						
	32 Deg.	40 Deg.	50 Deg.	60 Deg.	70 Deg.	80 Deg.	90 Deg.
60	2.39	1.71	9.86	.....	.....	.....	.....
80	4.09	3.43	2.59	1.74	0.88	.....	.....
100	5.79	5.14	4.32	3.49	2.64	1.77	.90
120	7.50	6.85	6.05	5.23	4.40	3.55	2.68
140	9.20	8.57	7.77	6.97	6.15	5.32	4.47
160	10.90	10.28	9.50	8.72	7.91	7.09	6.26
180	12.60	12.00	11.23	10.46	9.68	8.87	8.06
200	14.36	13.71	13.00	12.20	11.43	10.65	9.85
220	16.00	15.42	14.70	14.00	13.19	12.33	11.64

	100 Deg.	120 Deg.	140 Deg.	160 Deg.	180 Deg.	200 Deg.
60	.....	.....	.....	.....	.....	.....
80	.....	.....	.....	.....	.....	.....
100	.....	.....	.....	.....	.....	.....
120	1.80	.....	.....	.....	.....	.....
140	3.61	1.84	.....	.....	.....	.....
160	5.42	3.67	1.87	.....	.....	.....
180	7.23	5.52	3.75	1.91	.....	.....
200	9.03	7.36	5.62	3.82	1.96	.....
220	10.84	9.20	7.50	5.73	3.93	1.98

A good feed-water heater of adequate proportions should readily raise the temperature of feed-water up to 200° Fahr., and, as is seen by inspection of the table, thus effect a saving of fuel, ranging from 14.3 per cent. to 9.03 per cent., according as the atmospheric or normal temperature of the water varies from 32° Fahr. in the height of winter, to 100° Fahr. in the height of summer.

POINTS RELATING TO FEED WATER HEATERS.

The percentage of saving which may be obtained from the use of exhaust steam for heating the feed water, with which the boiler is supplied, will depend upon the temperature to which the water is raised, and this, in turn, will depend upon the length of time that the water remains under the influence of the exhaust steam. This should be as long as possible, and unless a sufficient amount of heating surface is employed in the heater best results cannot be expected.

It does not necessarily require all the exhaust steam—or the whole volume of waste steam passing from the engine to bring the feed water up to the temperature desired, and the larger the heating appliance the smaller proportion is needed—hence heaters are best made with two exits nicely proportioned to avoid back pressure and at the same time utilize enough of the exhaust to heat the feed water.

An impression prevails among many who are running a condenser on their engine that a feed water heater can not be used in connection with it; large numbers of heaters running on condensing engines with results as follows: the feed water is delivered to the boiler at a temperature of 150° to 160° Fahr., depending on the vacuum: the higher the vacuum the less the heat in the feed water.

A heater applied to a condensing engine generally increases the vacuum one to two inches.

When cold water is used for the feed water, the saving in fuel by the use of the heater is from 7 to 14 per cent.

When feed water is taken from the hot well, it will save 7 to 8 per cent.

Where all the steam generated by a boiler is used in the engine and the exhaust passed through a heater it is found by actual experiment, where iron tubes are used in the heater, that approximately ten square feet of heating surface will be required for each 30 lbs. of water supplied to the boiler at a temperature of 200 degrees Fahr.

Ten square feet of heating surface in the feed water heater also represents one horse power.

CAPACITY OF CISTERNS.

The following table gives the capacity of cisterns for each twelve inches in depth:

Diameter.	*Gallons.*
25 feet	3671
20 "	2349
15 "	1321
14 "	1150
13 "	992
12 "	846
11 "	710
10 "	587
9 "	475
8 "	376
7 "	287
6½ "	247
6 "	211
5 "	147
4½ "	119
4 "	94
3 "	53
2½ "	36
2 "	23

Supposing it was required to find the weight of the water in any cistern or tank; it can be ascertained by multiplying the number of gallons by the weight of one gallon, which is 8⅓ pounds, 8.333. For instance, taking the largest cistern in the above table containing 3671 gallons: 3671×8.33=30579.43 pounds.

The table above gives the capacities of round cisterns or tanks. If the cistern is rectangular the number of gallons and weight of water are found by multiplying the dimensions of the cistern to get the cubical contents. For instance, for a cistern or tank 96 inches long, 72 inches wide, and 48 inches deep, the formula would be: 96×72×48=331,776 cubic inches.

As a gallon contains 231 cubic inches; 331,776 divided by 231 gives 1,436 gallons, which multiplied by 8.33 will give the weight of water in the cistern.

CAPACITY OF CISTERNS.

For round cisterns or tanks, the rule is: Area of bottom on inside multiplied by the height, equals cubical capacity. For instance, taking the last tank or cistern in the table: Area of 24 inches (diameter) is 452.39, which multiplied by 12 inches (height) gives 5427.6 cubic inches, and this divided by 231 cubic inches in a gallon gives 23 gallons.

Supposing the tank to be 24 inches deep instead of 12 inches, the result would be, of course, twice the number of gallons.

RULE FOR OBTAINING CONTENTS OF A BARREL IN GALLONS.

Take diameter at bung, then square it, double it, then add square of head diameter; multiply this sum by length of cask, and that product by .2618 which will give volume in cubio inches; this, divided by 231, will give result in gallons.

WATER METERS.

Water meters, or measurers (apparatus for the measurement of water), are constructed upon two general principles: 1, an arrangement called an *"inferential meter"* made to divert a certain proportion of the water passing in the main pipe and by measuring accurately the small stream diverted, *to infer*, or estimate the larger quantity; 2, *the positive meter;* rotary piston meters are of the latter class and the form usually found in connection with steam plants. They are constructed on the positive displacement principle, and have only one working part—a hard rubber rolling piston—rendering it almost, if not entirely, exempt from liability to derangement. It measures equally well on all sized openings, whether the pressure be small or great; and its piston, being perfectly balanced, is almost frictionless in its operation.

Constructed of composition (gun-metal) and hard rubber, it is not liable to corrosion. An ingenious stuffing-box insures at all times a perfectly dry and legible dial, or the registering

WATER METERS.

mechanism which is made of a combination of metals especially chosen for durability and wear, and inclosed in a case of gun-metal.

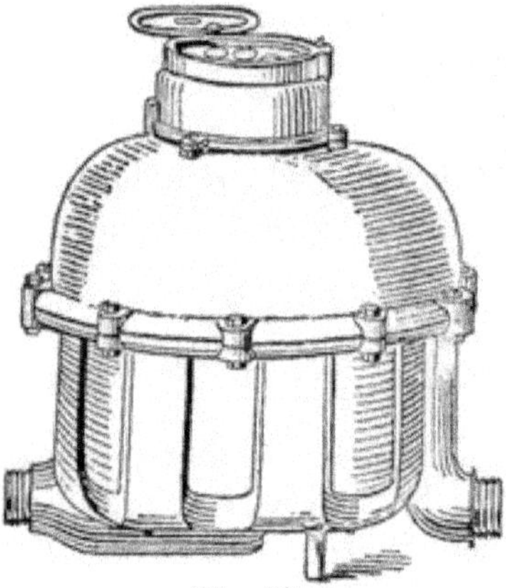

Fig. 99.

Fig. 99 is a perspective view of the meter, showing the index on the top. It is shown here as when placed in position. The proper threads at the inlet and outlet make it easy of attachment to the supply and discharge pipes.

The hard rubber piston (the only working part of the Meter), is made with spindle for moving the lever communicating with the intermediate gear by which the dial is moved.

The water, through the continuous movement of the piston, passes through the meter in an unbroken stream, in the same quantity as with the pipe to which it is attached when the opening in the meter equals that of the service pipe; the apparatus is noiseless and practically without essential wear.

"POINTS" RELATING TO WATER METERS.

In setting a meter in position let it be plumb, and properly secured to remain so. It should be well protected from frost.

If used in connection with a steam boiler, or under any other conditions where it is exposed to a back pressure of steam or hot water it must be protected by a check valve, placed between the outlet of the meter and the vessel it supplies.

It is absolutely necessary to blow out the supply pipe before setting a new meter, so that if there be any accumulation of sand, gravel, etc., in it, the same may be expelled, and thus prevented from entering the meter. Avoid using red lead in making joints. It is liable to work into the meter and cause much annoyance by clogging the piston.

WATER METERS.

This engraving, Fig. 100, shows the counter of the Meter. It registers cubic feet—one cubic foot being $7\frac{48}{100}$ U. S. gallons and is read in the same way as the counters of gas meters.

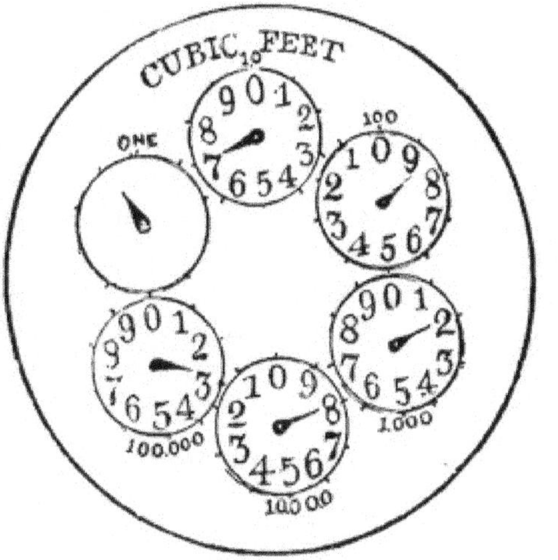

Fig. 100.

The following example and directions may be of service to those unacquainted with the method:

If a pointer be between two figures, the smallest one must always be taken. When the pointer is so near a figure that it seems to indicate that figure exactly, look at the dial next below it in number, and if the pointer there has passed 0, then the count should be read for that figure. Let it be supposed that the pointers stand as in the above engraving, they then read 28,187 cubic feet. The figures are omitted from the dial marked "ONE," because they represent but tenths of one cubic foot, and hence are unimportant. From dial marked "10," we get 7; from the next marked "100," we get 8; from the next marked "1,000," we get the figure 1; from the next marked "10,000," the figure 8; from the next marked "100,000," the figure 2.

THE FISH TRAP used in connection with water meters is an apparatus (as its name denotes) for holding back fishes, etc.

THE STEAM BOILER INJECTOR.

For safety sake, every boiler ought to have two feeds in order to avoid accidents when one of them gets out of order, and one of these should be an injector.

This consists in its most simple form, of a steam nozzle, the end of which extends somewhat into the second nozzle, called the combining or suction nozzle; this connects with, or rather terminates in, a third nozzle or tube, termed the "forcer." At the end of the *combining tube,* and before entering the forcer, is an opening connecting the interior of the nozzle at this point with the surrounding area. This area is connected with the outside air by a check valve, opening outward in the automatic injectors, and by a valve termed the overflow valve.

The operation of the injector is based on the fact, first demonstrated by Gifford, that the motion imparted by a jet of steam to a surrounding column of water is sufficient to force it into the boiler from which the steam was taken, and, indeed, into a boiler working at a higher pressure. The steam escaping from under pressure has, in fact, a much higher velocity than water would have under the same pressure and condition. The rate of speed at which steam—taking it at an average boiler pressure of sixty pounds—travels when discharged into the atmosphere, is about 1,700 feet per second. When discharged with the full velocity developed by the boiler pressure through a pipe, say an inch in diameter, the steam encounters the water in the combining chamber. It is immediately condensed and its bulk will be reduced say 1,000 times, but its velocity remains practically undiminished. Uniting with the body of water in the combining tube, it imparts to it a large share of its speed, and the body of water thus set in motion, operating against a comparatively small area of boiler pressure, is able to overcome it and pass into the boiler. The weight of the water to which steam imparts its velocity gives it a momentum that is greater in the small area in which its force is exerted than the boiler pressure, although its force has actually been derived from the boiler pressure itself.

THE STEAM INJECTOR.

The following cut 101 represents the outline of one of the best of a large number of injectors upon the market, from which the operation of injectors may be illustrated.

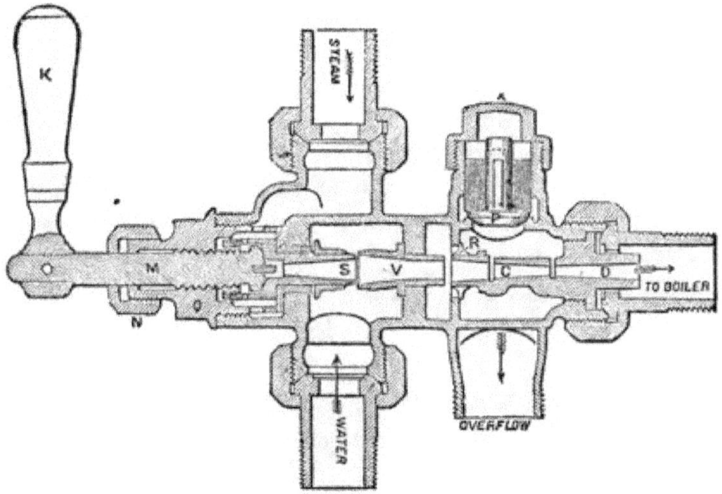

S. Steam jet. V. Suction jet. C-D. Combining and delivery tube. R. Ring or auxiliary check. P. Overflow valve. O. Steam plug. M. Steam valve and stem. N. Packing nut. K. Steam valve handle X Overflow cap

Fig. 101.

The steam enters from above, the flow being regulated by the handle K. The steam passes through the tube S and expands in the tube V, where it meets the water coming from the suction pipe. The condensation takes place in the tubes V and C, and a jet of water is delivered through the forcer tube D to the boiler. Connection passages are made to the chamber surrounding the tubes C, D, and to the end of tube V. If the pressure in this surrounding chamber becomes greater than that of the atmosphere, the check valve P is lifted and the contents are discharged through the overflow.

So long as the pressure in this chamber is atmospheric, the check valve P remains closed, and all the contents must be discharged through the tube D.

THE STEAM INJECTOR.

There are three distinct types of live steam injectors, the "simple fixed nozzle," the "adjustable nozzle," and the "double." The first has one steam and one water nozzle which are fixed in position but are so proportioned as to yield a good result. There is a steam pressure for every instrument of this type at which it will give a maximum delivery, greater than the maximum delivery for any other steam pressure either higher or lower. The second type has but one set of nozzles, but they can be so adjusted relative to each other as to produce the best results throughout a long range of action; that is to say, it so adjusts itself that its maximum delivery continually increases with the increase of steam pressure.

The double injector makes use of two sets of nozzles, the "lifter" and "forcer." The lifter draws the water from the reservoir and delivers it to the forcer, which sends it into the boiler. All double injectors are fixed nozzle.

All injectors are similar in their operation. They are designed to bring a jet of live steam from the boiler in contact with a jet of water so as to cause it to flow continuously in the direction followed by the steam, the velocity of which it in part assumes, back into the boiler and against its own pressure.

As a thermodynamical machine, the injector is nearly perfect. since all the heat received by it is returned to the boiler, except such a very small part as may be lost by radiation; consequently its thermal efficiency should be in every case nearly 100 per cent. On the other hand, because of the fact that its heat energy is principally used in warming up the cold water as it enters the injector, its mechanical efficiency, or work done in lifting water, compared with the heat expended, is very low.

The action of the injector is as follows: Steam being turned on, it rushes with great velocity through the steam nozzle into and through the combining tube. This action induces a flow of air from the suction pipe, which is connected to the combining tube, with the result that a more or less perfect vacuum is formed, thus inducing a flow of water. After the water commences to flow to the injector it receives motion from the jet of steam: it absorbs heat from the steam and finally condenses it, and thereafter moves on into the forcer tube simply as a stream of water, at a low velocity compared with that of the steam. At the beginning of the forcer tube it is subjected only to atmospheric pressure, but from this point the pressure increases and the water moves forward at diminished velocity.

"Points" Relating to the Injector.

In nine cases out of ten, where the injector fails to do good service, it will be either because of its improper treatment or location, or because too much is expected of it. The experience of thoroughly competent engineers establishes the fact that in almost every instance in which a reliable boiler feed is required, an injector can be found to do the work, provided proper care is exercised in its selection.

The exhaust steam injector is a type different from any of the above-named, in that it uses the exhaust steam from a non-condensing engine. Exhaust steam has fourteen and seven-tenths (14.7) pounds of work, and the steam entering the injector is condensed and the water forced into the boiler upon the same general principle as in all injectors.

The exhaust steam injector would be still more extensively used were it not for a practical objection which has arisen—it carries over into the boiler the waste oil of the steam cylinder.

Some injectors are called by special names by their makers, such as ejecters and inspirators, but the term injectors is the general name covering the principle upon which all the devices act.

The injector can be and sometimes is, used as a pump to raise water from one level to another. It has been used as an air compressor, and also for receiving the exhaust from a steam engine, taking the place in that case of both condenser and air pump.

The injector nozzles are tubes, with ends rounded to receive and deliver the fluids with the least possible loss by friction and eddies.

Double injectors are those in which the delivery from one injector is made the supply of a second, and they will handle water at a somewhat higher temperature than single ones with fixed nozzles.

POINTS RELATING TO THE INJECTOR.

The motive force of the injector is found in the heat received from the steam. The steam is condensed and surrenders its latent heat and some of its sensible heat. The energy so given up by each pound of steam amounts to about 900 thermal units, each of which is equivalent to a mechanical force of 778 foot pounds. This would be sufficient to raise a great many pounds of water against a very great pressure could it be so applied, but a large portion of it is used simply to heat the water raised by the injector.

The above explanation will apply to every injector in the market, but ingenious modifications of the principles of construction have been devised in order to meet a variety of requirements.

That the condensation of the steam is necessary to complete the process will be evident, for if the steam were not condensed in the combining chamber, it would remain a light body and, though moving at high speed, would have a low degree of energy.

Certain injectors will not work well when the steam pressure is too high. In order to work at all the injector must condense the steam which flows into the combining tube. Therefore, when the steam pressure is too high, and as a consequence the heat is very great, it is difficult to secure complete condensation; so that for high pressure of steam good results can only be obtained with cold water. It would be well when the feed water is too warm to permit the injector to work well, to reduce the pressure, and consequently the temperature of the steam supplied to the injector, as low pressure steam condenses much easier, and consequently can be employed with better result. Throttling the steam supplied by means of stop valves will often answer well in this case. The steam should not be cold or it will not contain heat units enough to allow it to condense into a cross section small enough to be driven into the boiler. This is the reason why exhaust injectors fail to work when the exhaust steam is very cold. It also explains why such injectors work well when a little live steam is admitted into the exhaust sufficient to heat it above a temperature of 212°.

POINTS RELATING TO THE INJECTOR.

Leaks affect injectors the same as pumps, and in addition, the accumulation of lime and other mineral deposits in the jets stops the free flowing of the water. The heat of the steam is the usual cause of the deposits, and where this is excessive it would be well to discard the injector and feed with the pump.

The efficient working of the injector depends materially upon the size of the jet which should be left as the manufacturer makes it; hence in repairs and cleaning a scraper or file should not be used.

For cleaning injectors, where the jets have become scaled, use a solution of one part muriatic acid to from nine to twelve parts of water. Allow the tubes to remain in the acid until the scale is dissolved or is so soft as to wash out readily.

The lifting attachment, as applied to any injector, is simply a steam jet pump. It is combined with the injector proper and is operated by a portion of the steam admitted to the instrument. Nearly all the successful injectors on the market are made with these attachments, and will raise water about 25 feet if required, from a well or tank below the boiler level.

Where an injector is required to work at different pressures it must be so constructed that the space between the receiving tube and the combining tube can be varied in size. As a rule this is accomplished by making both combining and receiving tubes conical in form and arranging the combining tube so that it can be moved to or from the receiving tube, and the water space thereby enlarged or contracted at will. The adjustment of the space between the two tubes by hand is a matter of some difficulty, however; at least it takes more time and patience than the average engineer has to devote to it, and the majority of the injectors in use are therefore made automatic in their regulation.

The injector is not an economical device, but it is simple and convenient, it occupies but a small amount of space, is not expensive and is free from severe strains on its durability; moreover, where a number of boilers are used in one establishment, it is very convenient to have the feeding arrangements separate, so that each boiler is a complete generating system in itself and independent ot its neighbors.

LAWS OF HEAT.

Heat is a word freely used, yet difficult to define. The word "heat" is commonly used in two senses: (1) to express the sensation of warmth; (2) the state of things in bodies which causes that sensation. The expression herein must be taken in the latter sense.

Heat is transmitted in three ways—by *conduction*, as when the end of a short rod of iron is placed in a fire, and the opposite end becomes warmed – this is conducted heat; by *convection* (means of currents). such as the warming of a mass of water in a boiler, furnace, or saucepan; and by *radiation*, as that diffused from a piece of hot metal or an open fire. Radiant heat is transmitted, like sound or light, in straight lines in every direction, and its intensity diminishes inversely as the square of the distance from its center or point of radiation. Suppose the distance from the center of radiation to be 1, 2, 3 and 4 yards, the surface covered by heat rays will increase 1, 4, 9 and 16 square feet; the intensity of heat will diminish 1, $\frac{1}{4}$, 1-9, and 1-16. and so on in like proportions, until the heat becomes absorbed, or its source of supply stopped.

Whenever a difference in temperature exists, either in solids or liquids that come in contact with or in close proximity to each other, there is a tendency for the temperature to become equalized; if water at 100° be poured into a vessel containing an equal quantity of water at 50°, the tendency will be for the whole to assume a temperature of 75°; and suppose the temperature of the surrounding air be 30°, the cooling process will continue until the water and the surrounding air become nearly equal, the temperature of the air being increased in proportion as that of the water is decreased.

The heat generated by a fire under the boiler is transmitted to the water inside the boiler, when the difference in the specific gravities, or, in other words, the cold water in the pipes being heavier than that in the boiler sinks and forces the lighter hot water upward. This heat is radiated from the pipes, which are good conductors of heat to the air in the room, and raises it to the required temperature. That which absorbs heat

LAWS OF HEAT.

rapidly, and parts with it rapidly, is called a good conductor, and that which is slow to receive heat, and parts with it slowly, is termed a bad conductor.

The following tables of conductivity, and of the radiating properties of various materials, may be of service:

CONDUCTING POWER OF VARIOUS SUBSTANCES.—DESPRTIZ.

Material.	*Conductivity.*
Gold	100
Silver	97
Copper	89
Brass	75
Cast iron	56
Wrought iron	37
Zinc	36
Tin	30
Lead	18
Marble	2.4
Fire clay	1.1
Water	0.9

RADIATING POWER OF VARIOUS SUBSTANCES.—LESLIE

Material.	*Radiating Power.*
Lampblack	100
Water	100
Writing paper	98
Glass	90
Tissue paper	88
Ice	85
Wrought lead	45
Mercury	20
Polished lead	19
Polished iron	15
Gold, silver	12
Copper, tin	12

From the above tables, it will be seen that water, being an excellent radiator, and of great specific heat, and iron a good conductor, these qualities, together with the small cost of the materials, combine to render them efficient, economic and convenient for the transmission and distribution of artificial heat.

LAWS OF HEAT.

By adopting certain standards we are enabled to define, compare and calculate so as to arrive at definite results, hence the adoption of a standard unit of heat, unit of power, unit of work, etc.

The standard unit of heat is the amount necessary to raise the temperature of one pound of water at 32° Fahr. on degree, *i. e.*, from 32° to 33°.

Specific heat is the amount of heat necessary to raise the temperature of a solid or liquid body a certain number of degrees; water is adopted as the unit or standard of comparison. The heat necessary to raise one pound of water one degree, will raise one pound of mercury about 30 degrees, and one pound of lead about 32 degrees.

TABLE OF THE SPECIFIC HEAT OF EQUAL WEIGHTS OF VARIOUS SUBSTANCES.

Solid bodies.	*Specific Heat.*
Wood (fir and pine)	0.650
" (oak)	0.570
Ice	0.504
Coal	0.280
Charcoal (animal)	0.260
" (vegetable)	0.241
Iron (cast)	0.241
Coke	0.201
Limestone	0.200
Glass	0.195
Steel (hard)	0.117
" (soft)	0.116
Iron (wrought)	0.111
Zinc	0.095
Copper (annealed)	0.094
" (cold hammered)	0.093
Tin	0.056
Lead	0.031
Liquids.	
Water	1.000
Alcohol	0.158
Acid (pyroligneous)	0.590
Ether	0.520
Acid (acetic)	0.509
Oil (olive)	0.309
Mercury	0.033

LAWS OF HEAT.

Gases.

Hydrogen	3.409
Vapor of alcohol	0.547
Steam	0.480
Carbonic oxide	0.245
Nitrogen	0.243
Oxygen	0.217
Atmospheric air	0.237
Carbonic acid	0.202

THE STEAM PUMP.

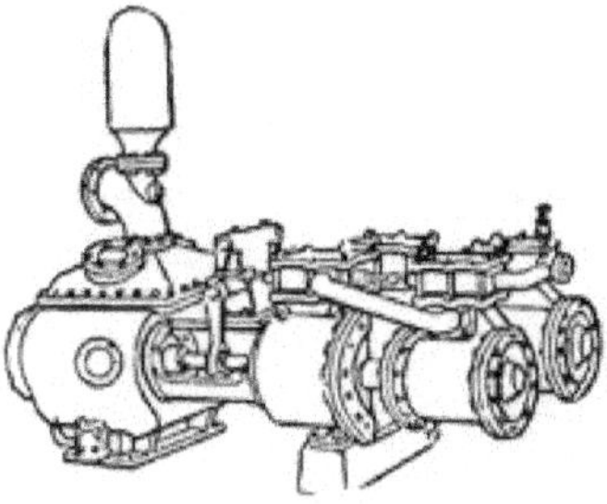

Fig. 102.

It is difficult to overestimate the importance, in connection with a steam plant, of the appliance which supplies water for the boiler, not only, but a hundred other uses. Upon the steady operation of the pump depends the safety and comfort of the engineer, owner and employee, and indirectly of the success of the business with which the "plant" is connected. Hence the necessity of acquiring complete knowledge of the operation of a device so important.

Pumps now raise, convey and deliver water, beer, molasses, acids, oils, melted lead. Pumps also handle, among the gases, air, ammonia, lighting gas, and oxygen. Pumps are also used to increase or decrease the pressure of a fluid.

Pumps are made in many ways, and defined as rope, chain, diaphram, jet, centrifugal, rotary, oscillating, cylinder.

Cylinder pumps are of two classes, single acting and double acting. In single acting—in effect is *single ended*—in double acting, the motion of the cylinder in one direction causes an inflow of water and a discharge at the same time, in the other; and on the return stroke the action is renewed as the discharge end becomes the suction end. The pump is thus double acting.

STEAM PUMPS.

A *direct pressure* steam pump is one in which the liquid is pressed out by the action of steam upon its surface, without the intervention of a piston. A direct acting steam pump is an engine and pump combined.

A cylinder or reciprocating pump is one in which the piston or plunger, in one direction, causes a partial vacuum, to fill which the water rushes in pressed by the air on its head.

NOTE.—A *suction valve* prevents the return of this water on the return stroke of the piston and a *discharge valve* permits the outward passage of the fluid from the pump but not its return thereto or to the reservoir through the suction pipe.

The force against which the pump works is gravity or the attraction of the earth which prevents the water from being lifted. This is shown in the fact that water can be led. or trailed, an immense distance, limited only by the friction, by a pump.

NOTE.—It may be noted that the difference between a fluid and *liquid* is shown in the fact that the latter can be poured from one vessel to another, thus: air and water are both fluids, but of the two water alone is liquid : air, ammonia, etc., are *gases,* while they are also fluids, *i. e.,* they flow

The idea entertained by many that water is raised by suction, is erroneous. Water or other liquids are raised through a tube or hose by the pressure of the atmosphere on their surface. When the atmosphere is removed from the tube there will be no resistance to prevent the water from rising, as the water outside the pipe, still having the pressure of the atmosphere upon its surface, forces water up into the pipe, supplying the place of the excluded air, while the water inside the pipe will rise above the level of that outside of it proportionally to the extent to which it is relieved of the pressure of the air.

If the first stroke of a pump reduces the pressure of the air in the pipe from 15 pounds on the square inch to 14 pounds, the water will be forced up the pipe to the distance of 2¼ feet, since a column of water an inch square and 2¼ feet high is equal in weight to about 1 pound. Now if the second stroke of the pump reduces the pressure of the atmosphere in the pipe to 13 pounds per inch, the water will rise another 2¼ feet; this rule is uniform, and shows that the rise of the column of water within the pipe is equal in weight to the pressure of the air upon the surface of the water without.

There are pumps (Centrifugal) especially designed for pumping water mingled with mud, sand, gravel, shells, stones, coal, etc., but with these the engineer has but little to do, as they are used mostly for wrecking and drainage.

The variety of pattern in which pumps are manufactured and the still greater variation in capacity forbids an attempt to fully illustrate and describe further than their general principles, and to name the following general

Classification of Pumps.

1st. Pumps are divided into Vertical and Horizontal.

Vertical Pumps are again divided into:

1. Ordinary Suction or Bucket Pumps.
2. Suction and Lift Pumps.
3. Plunger or Force Pumps.
4. Bucket and Plunger Pumps.
5. Piston and Plunger Pumps.

Horizontal Pumps are divided into:

1. Double-acting Piston Pumps.
2. Single-acting Plunger Pumps.
3. Double-acting Plunger Pumps.
4. Bucket and Plunger Pumps.
5. Piston and Plunger Pumps.

Fig. 103.

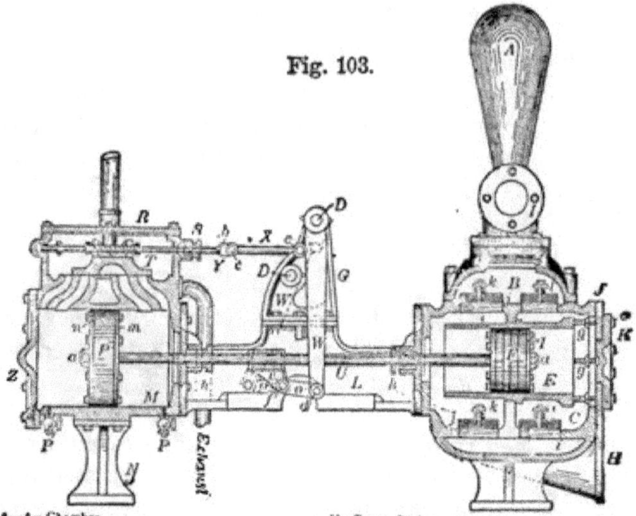

A—Air Chamber.
B—Water Cylinder Cap.
C—Water Cylinder with Valves and Seats in.
D—Rocker Shafts, each, Long or Short
E—Removable Cylinders, each.
F—Water Piston and Follower, each.
g—Water Piston Followers, each.
G—Rocker Stand.
H—Suction Flange, threaded.
I—Discharge Flange, threaded.
J—Intermediate Flanges, each.
K—Water Cylinder Heads, each.
L—Concaves complete, with Stuffing Boxes, each
M—Steam Cylinder, without Head, Bonnet and Valve.
N—Steam Cylinder Foot.
O—Crosshead Links, each.
P—Steam Piston, complete with Rings and Follower, each.
m—Steam Piston Head.
n—Steam Piston Follower.
Steam Piston Rings, including Spring and Breakjoint.
Q—Side Water Cylinder Bonnet, each.
R—Steam Chest Bonnet, each
S—Steam Chest Stuffing Box Gland, each.
T—Steam Slide Valve, each.
U—Piston Rods, each.
V—Cross-heads, each.
W—Rocker Arms, each, Long or Short.
X—Valve Rod Links, each, Long or Short.
Y—Steam Valve Stems, each.
Z—Steam Cylinder Heads, each.
aa—Piston Rod Nuts, each.
hh—Piston Rod Stuffing Glands, each.
ii—Water Valve Seats, each.
jj—Rubber Valves, each
kk—Water Valve Stems, each.
ll—Water Valve Springs, each.
gg—Removable Cylinder Screws, each.
b—Steam Valve Stem Forks, each.
c—Steam Valve Stem Fork Bolts, each.
e—Valve Rod Link Bolts, each.
d—Rocker Arm Pins, each.
f—Crosshead Link Bolts, each.
o—Collar Bolts, each.
pp—Brass Steam Cylinder Drain Cocks, each.
Water Packings, each.
Brass Piston Rods, each.
Brass Lined Removable Cylinders, extra, each.
Piston Rod Stuffing Gland Bolts, each.
Water Cylinder Cap Bonnets, each.
Top Valve Caps, each.
Valve Cap Clamps, each.

In Figs. 102 and 103 are exhibited the outlines of *the double acting steam pump*, which is undoubtedly the pattern most thoroughly adapted for feeding steam boilers, as it is equipped for the slowest motion with less risk of stopping on a center.

From the drawing with reference letters may be learned the terms applied generally to the parts of all steam pumps: example: "k" shows the water valve stems, "K" the water cylinder heads.

It may be remarked that nearly all pump makers furnish valuable printed matter, giving directions *as to repairs*, and best method of using their particular pumps—especially valuable are their repair sheets in which are given cuts of "parts" of the pumps. It were well for the steam user and engineer to request such matter from the manufacturers for the special pump they use.

POINTS RELATING TO PUMPS.

Blow out the steam pipe thoroughly with steam before connecting it to the engine; otherwise any dirt or rubbish there might be in the pipe will be carried into the steam cylinder, and cut the valves and piston.

Never change the valve movement of the engine end of the pump. If any of the working parts become loose, bent or broken, replace them or insert new ones, in precisely the same position as before.

Keep the stuffing boxes nearly full of good packing well oiled, and set just tight enough to prevent leakage without excessive friction.

Use good oil only, and oil the steam end just before stopping the pump.

It is absolutely necessary to have a full supply of water to the pump.

If possible avoid the use of valves and elbows in the suction pipe, and see that it is as straight as possible; for bends, valves and elbows materially increase the friction of the water flowing into the pump.

See that the suction pipe is not imbedded in sand or mud, but is free and unobstructed.

All the pipes leading from the source of supply to the pump must be air-tight, for a very small air-leak will destroy the vacuum, the pump will not fill properly; its motion will be jerky and unsteady, and the engine will be liable to breakage.

A suction air chamber (made of a short nipple, a tee, a piece of pipe of a diameter not less than the suction pipe and from two to three feet long, and a cap, screwed upright into the suction pipe close to the pump) is always useful; and where the suction pipe is long, in high lifts, or when the pump is running at high speed, it is a positive necessity.

Never take a pump apart before using it. If at any time subsequently the pump should act badly, always examine the pump end first. And if there is any obstruction in the valve, remove it. See that the pump is well packed, and that there are no cracks in pipes or pump, nor any air-leaks.

POINTS RELATING TO PUMPS.

In selecting a pump for boiler feeding it is well to have it plenty large enough, and also these other desirable features: few parts, have no dead points or center, be quiet in operation, economical of steam and repairs, and positive under any pressure.

Granted motion to the piston or plunger, a pump fails because it leaks. There can be no other reason, and the leak should be found and repaired. Leaky valves are common and should be ground. Leaky pistons are not so common, but sometimes occur. Repairing is the remedy. Leaky plungers are common. They need re-turning. The rod must be straight as far as in contact with the packing. The packing around the plungers is sometimes neglected too long, gets filled with dirt and sediment, and hardens and scores an otherwise perfect rod, and so leaks.

The lifting capacity of a pump depends upon proper proportion of clearance in the cylinder and valve chamber, to displacement of the piston and plunger.

An injector is a sample of a *jet pump*—this may either lift or force or both.

The most necessary condition to the satisfactory working of the steam pump is a full and steady supply of water. The pipe connections should in no case be smaller than the openings in the pump. The suction lift and delivery pipes should be as straight and smooth on the inside as possible.

When the lift is high, or the suction long, a foot valve should be placed on the end of the suction pipe, and the area of the foot valve should exceed the area of the pipe.

The area of the steam and exhaust pipes should in all cases be fully as large as the nipples in the pump to which they are attached.

The distance that a pump will lift or draw water, as it is termed, is about 33 feet, because water of one inch area 33 feet weighs 14.7 pounds; but pumps must be in good order to lift 33 feet, and all pipes must be air-tight. Pumps will give better satisfaction lifting from 22 to 25 feet.

POINTS RELATING TO PUMPS.

In cold weather open all the cocks and drain plugs to prevent freezing when the pump is not in use.

When purchasing a steam pump to supply a steam boiler, one should be selected capable of delivering one cubic foot of water per horse-power per hour.

No pump, however good, will lift hot water, because as soon as the air is expelled from the barrel of the pump the vapor occupies the space, destroys the vacuum, and interferes with the supply of water. As a result of all this the pump knocks. When it becomes necessary to pump hot water, the pump should be placed below the supply, so that the water may flow into the valve chamber.

The air vessel on the delivery pipe of the steam pump should never be less than five times the area of the water cylinder.

There are many things to be considered in locating steam pumps, such as the source from which water is obtained, the point of delivery, and the quantity required in a given time; whether the water is to be lifted or flows to the pump; whether it is to be forced directly into the boiler, or raised into a tank 25, 50 or 100 feet above the pump.

The suction chamber is used to prevent pounding when the pump reverses and to enable the pump barrel to fill when the speed is high.

Suction is the unbalanced pressure of the air which is at sea level $14\frac{7}{10}$ per inch, or 2096.8 per square foot.

When a valve is spoken of in connection with a pump it may be understood that there may be several valves dividing and performing the functions of one.

A simple method of obtaining tight pump-valves consists simply in grooving the valve-sheets and inserting a rubber cord in the grooves. As the valves seat themselves the cord is compressed and forms a tight joint. An additional advantage is that it prevents the shock ordinarily produced by rapid closing and prolongs the life of the valve seat. The rubber cord when worn can be easily and quickly replaced.

CALCULATIONS RELATING TO PUMPS.

To find the pressure in pounds per square inch of a column of water, multiply the height of the column in feet by .434. Approximately, we say that every foot elevation is equal to ½ lb. pressure per square inch; this allows for ordinary friction.

To find the diameter of a pump cylinder to move a given quantity of water per minute (100 feet of piston being the standard of speed), divide the number of gallons by 4, then extract the square root, and the product will be the diameter in inches of the pump cylinder.

To find quantity of water elevated in one minute running at 100 feet of piston speed per minute. Square the diameter of the water cylinder in inches and multiply by 4. Example: capacity of a 5 inch cylinder is desired. The square of the diameter (5 inches) is 25, which, multiplied by 4, gives 100, the number of gallons per minute (approximately).

To find the horse power necessary to elevate water to a given height, multiply the weight of the water elevated per minute in lbs. by the height in feet, and divide the product by 33,000 (an allowance should be added for water friction, and a further allowance for loss in steam cylinder, say from 20 to 30 per cent.).

The area of the steam piston, multiplied by the steam pressure, gives the total amount of pressure that can be exerted. *The area of the water piston,* multiplied by the pressure of water per square inch, gives the resistance. *A margin* must be made between the *power* and the *resistance* to *move* the piston at the required speed—say from 20 to 40 per cent., according to speed and other conditions.

To find the capacity of a cylinder in gallons. Multiplying the area in inches by the length of stroke in inches will give the total number of cubic inches: divide this amount by 231 (which is the cubical contents of a U. S. gallon in inches), and product is the capacity in gallons.

The temperature 62° F. is the temperature of water used in calculating the specific gravity of bodies, with respect to the gravity or density of water as a basis, or as unity.

STEAM PUMPS.

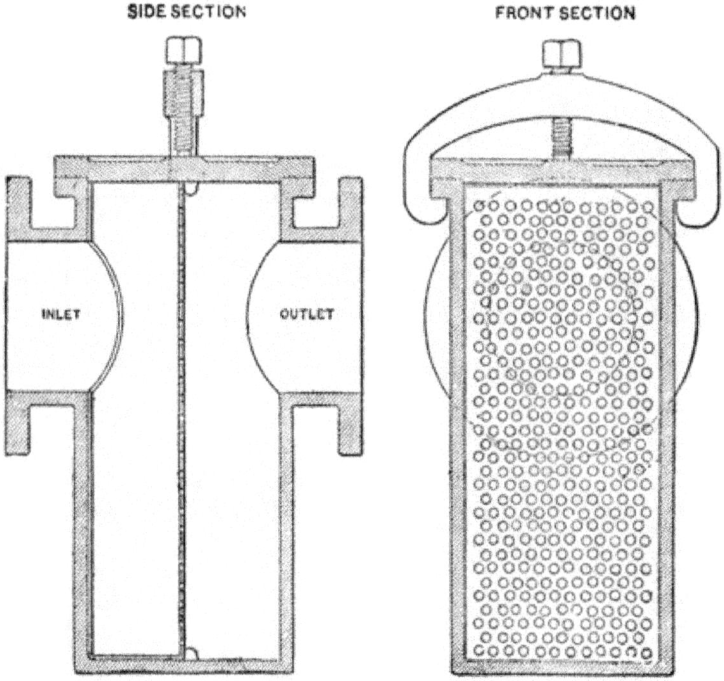

Fig. 104.

Important stress has been laid upon keeping all floating objects, gravel, etc., away from the acting parts of the pump. In Fig. 104 is presented a cut of an approved strainer which can be removed, freed from obstruction, and replaced by simply slacking one bolt, the entire operation occupying one minute. The advantages of this strainer will be readily apparent.

IMPORTANT PRINCIPLES RELATING TO WATER.

There are some underlying natural laws and other data relating to water which every engineer should thoroughly understand. Heat, *water*, steam, are the three properties with which he has first to deal.

IMPORTANT PRINCIPLES RELATING TO WATER.

Weight of one cubic foot of Pure Water.

At 32° F.		= 62.418 pounds.
At 39°.1		= 62.425 "
At 62°	(Standard temperature)	= 62.355 "
At 212°		= 59.640 "

The weight of a cubic foot of water is about 1000 ounces (exactly 998.8 ounces), at the temperature of maximum density.

The weight of a cylindrical foot of water at 62° F. is 49 lbs. (nearly). The weight of a cylindrical inch is 0.4533 oz.

There are four notable temperatures for water, namely,

32° F.,	or	0° C.	= the freezing point under one atmosphere.
39°.1	or	4°	= the point of maximum density.
62°	or	16°.66	= the standard temperature.
212°	or	100°	= the boiling point, under one atmosphere.

Water rises to the same level in the opposite arms of a recurved tube, hence water will rise in pipes as high as its source.

The pressure on any particle of water is proportioned to its depth below the surface, and as the side pressure is equal to the downward pressure.

Water at rest presses equally in all directions. This is a most remarkable property, the upward direction of the pressure of water is equal to that pressing downwards, and the side pressure is also equal.

Any quantity of water, however small, may be made to balance any quantity, however great. This is called the Hydrostatic Paradox, and is sometimes exemplified by pouring liquids into casks through long tubes inserted in the bung holes. As soon as the cask is full and the water rises in the pipe to a certain height the cask bursts with violence.

Water is practically non-elastic. A pressure has been applied of 30,000 pounds to the square inch and the contraction has been found to be less than one-twelfth.

POINTS RELATING TO WATER.

The surface of water at rest is horizontal. A familiar example of this may be noted in the fact that the water in a battery of boilers seeks a uniform level, no matter how much the cylinders may vary in size.

A given pressure or blow impressed on any portion of a mass of water confined in a vessel is distributed equally through all parts of the mass; for example' a plug forced inwards on a square inch of the surface of water, is suddenly communicated to every square inch of the vessel's surface, however large, and to every inch of the surface of any body immersed in it.

WEIGHT AND CAPACITY OF DIFFERENT STANDARD GALLONS OF WATER.

	Cubic inches in a Gallon.	Weight of a Gallon in pounds.	Gallons in a cubic foot.	Weight of a cubic foot of water, English standard, 62.321 lbs. Avoirdupois.
Imperial or English.	277.274	10.00	6.232102	
United States	231.	8.33111	7.480519	

STORING AND HANDLING OF COAL.

The best method of storing coal is a matter of economy and needs the attention of the engineer.

Coal, as it comes from the mine, is in the best possible condition for burning in a furnace; its fracture is bright and clean, and it ought to be preserved up to the time of using it in such manner as to avoid as much as possible any alteration of its condition so as to prevent deterioration.

So far as actual experience goes it has been found that a brick buiding, with double walls to promote coolness, with high narrow slits instead of windows, with ventilating holes along the bottom of the walls, having a high-pitched roof with overhanging eaves, and holes for ventilation well sheltered under the eaves, and with ventilators along the edge of the roof, is best suited to keep the coal in the condition most nearly approaching that of the freshly mined. The floor of the building should be preferably paved with brick on edge or flagstones; the doors should be large and kept open in damp weather, and closed when the weather is hot.

Some persons recommend sprinkling the coal occasionally during the hot weather, but it is much better to wet down the paving all around the building outside, and the exposed floor of the building, as well as the walls inside and outside, and let the moisture of the evaporation have its effect upon the coal. It will be found to be amply sufficient for the purpose.

It has been found long since that it is better to have coal sheds dark, as light assists greatly in impairing the fuel.

The best arrangement for a boiler room floor is to have a coal-bin, paved with stone flags, opening into the fire-room by a door, while the fire-room itself should be paved diagonally with brick, set on edge upon a concrete foundation, well rammed to within about three feet of the boiler front, and the remaining space should be floored with iron plates.

The coal should be wheeled from the bins and dumped upon these plates, never on the brick floor. These plates should be laid on an incline of about an inch toward the boilers, and it is well to have a trough or gutter, of about six inches in wid h, and having a depth of about one and a half inches cast in them, at the edge lying nearest the boilers, so tnat the water from the gauge-cock, drip-pipes, and that from wetting down the ashes may run into it and drain into a proper sewer-pipe laid under the flooring.

CHEMISTRY OF THE FURNACE.

A careful estimase by a Broadway Chemist of the contents or constituents of a ton of coal presents some interesting facts, not familiar certainly to unscientific minds. It is found that, besides gas, a ton of ordinary gas coal will yield 1,500 pounds of coke, twenty gallons of ammonia water and 140 pounds of coal tar. Now, destructive distillation of this amount of coal tar gives about seventy pounds of pitch, seventeen pounds of creosote, fourteen pounds of heavy oils, about nine and a half

CHEMISTRY OF THE FURNACE.

pounds of naphtha yellow, six and one-third pounds of naphthaline, four and three-fourth pounds of alizarine, two and a fourth pounds of solvent naphtha, one and a fifth pound of aniline, seventy-nine hundredths of a pound of toludine, forty-six hundredts of a pound of anthracine, and nine-tenths of a pound of toluches—from the last-named substance being obtained the new product, saccharine, said to be 230 times as sweet as the best cane sugar.

From an engineer's standpoint the main constituents of all coal are carbon and hydrogen; in the natural state of coal these two are united and solid; their respective characters and modes of entering into combustion, are however essentially different. The hydrogen is convertable into heat only in the gaseous state; the carbon, on the contrary, is combustible only in the solid condition. It must be borne in mind that neither is combustible while they are united.

There are, however, other elements existing in coal in its natural state, and new ones are formed during burning or combustion as will be noted in the succeeding paragraphs.

For raising steam the process of combustion consists in disentangling, letting loose or evolving the different elements locked up in coal; the power employed in accomplishing this is *heat*. The chemical results of this consumption of the fuels may be divided into four stages or parts.

First stage, application of existing heat to disengage the constituent gases of the fuel. In coals this is principally mixed carbon and hydrogen.

Second stage, application or employment of existing heat to separate the carbon from the hydrogen.

Third stage, further employment of existing heat to increase the temperature of the two combustibles, carbon and hydrogen, until they reach the heat necessary for combination with the air. If this heat is not obtained, chemical union does not take place and the combustion is imperfect.

CHEMISTRY OF THE FURNACE.

Fourth and last stage, the union of the oxygen of the air with the carbon and hydrogen of the furnace in their proper proportions, when intense heat is generated and light is also given off from the ignited carbon. The temperature of the products of combustion at this final stage depend upon the quantity of air in dilution. Sir H. Davy estimates this heat as greater than the white heat of metals.

In the first stages heat is absorbed, but is given out in the last. When the chemical atoms of heat are not united in their proper proportions, then carbonic oxide, mixed carbon and hydrogen, and other combustible gases escape invisibly, with a corresponding loss of heat from the fuel.

When the proper union takes place, then only steam, carbonic acid and nitrogen, all of which are incombustible, escape.

The principal products, therefore, of perfect combustion are: steam, invisible and incombustible; carbonic acid, invisible and incombustible.

The products of imperfect combustion are: carbonic oxide, invisible but combustible; smoke, partly invisible and partly incombustible.

Steam is formed from the hydrogen gas given out by the coals combining with its equivalent of oxygen from the air. Smoke is formed from the hydrogen and carbon which have not received their respective equivalents of oxygen from the air, and thus pass off unconsumed. The color of the smoke depends upon the carbon passing off in its dark, powdery state.

The heat lost is not dependent upon the amount of carbon alone, but also upon the invisible but combustible gases, hydrogen and carbonic oxide; so that while the color may indicate the amount of carbon in the smake, it does not indicate the amount of the heat lost; hence, the smokeless locomotive burning coke may lose more heat in this way than that arising from the imperfect burning of coal under the stationary engine boiler.

CHEMISTRY OF THE FURNACE.

A practical and familiar instance of imperfect combustion is exhibited when a lamp smokes and the unconsumed carbon is deposited all about in the form of soot. When the evolving or disengagement of the carbon is reduced by lowering the wick to meet the supply of oxygen, the carbon is all consumed and the smoke ceases. What takes place in a lamp also occurs in a furnace, so that the proper supply of air is a primary thing, relating to economy, both as regards its quantity and its mode of admission to a fire.

The economical generation of heat is one thing, the use made of that heat afterwards is another. Combustion may be perfect, but the absorption of heat by a boiler may be inferior.

The chief agents operating in the furnace are carbon, hydrogen and oxygen, and their union in certain proportions produces other bodies, as water or steam, carbonic acid, besides others of less practical importance.

OXYGEN is an invisible gas, has no smell, and remains permanently in receptacles, unchanged by time. It can be obtained in an experimental quantity by heating the chlorate of potash, and collecting the gas given off in a bladder or jar. It is a trifle heavier than common air, *i. e.*,. 1.106 times and a cubic foot at 32° temperature weighs 1.428 ounces. It is one of the most abundant bodies in nature, and is combined with many others in a great variety of ways.

CARBON is one of the most interesting elementary substances in nature. It is combustible and forms the base of charcoal, and enters largely into mineral coal. It is a mineral capable of being reduced to a feathery powder, and is found in many different forms. It is obtained by various processes: from oil lamps as lamp-black; from coal as coke, and from wood as charcoal; the mineral particles of carbon in a state of combustion render flame luminous from either gas, oil or candles.

Carbon unites with iron to form steel, and with hydrogen to form the common street gas. Carbon is considered as the next most abundant body in nature to oxygen. In the furnace the carbon of the fuel unites with the oxygen of the air to produce heat; if the supply of air is correctly regulated, there will be perfect combustion, but if the supply of air be deficient, combustion will be imperfect.

HYDROGEN is an invisible gas, and the lightest known body in the world, being many times lighter than oxygen. It is combustible and gives out much heat. In our gas establishments it is made in large quantities and combined with carbon for illuminating streets, shops and dwellings. It is the source of all common flame. When united with sulphur in coal mines it becomes explosive. By passing a current of steam through a hot iron tube partly filled with filings, hydrogen gas is given off and burns with a pale yellow flame.

The more hydrogen, therefore, there is in the fuel, the greater in general is its heating power. But it must be borne in mind that the element of hydrogen is, nevertheless, to a greater or less degree neutralized by the other element, oxygen, when it is present as a constituent of the fuel; since the affinity of hydrogen for oxygen is superior to that of carbon, and the oxygen saturated with hydrogen is converted into steam and rises in this form from the fuel bed without producing heat. Thus it is that the more oxygen there is in the fuel the less is its power for developing heat by combustion.

NITROGEN is also an elementary body. It neither supports life nor combustion; it is lighter than air and has no taste or smell. One cubic foot at 32° temperature weighs a trifle less than one ounce.

SULPHUR is also an elementary body, of a yellow color, brittle, does not dissolve in water, is easily melted, and inflammable. It is also called brimstone or *burnstone*, from its great combustibility. It burns with a blue flame, and with a peculiar, suffocating odor.

CARBONIC ACID GAS is formed by the burning of sixteen parts of oxygen and six parts of carbon. Its specific gravity is 1.529; it is fatal to life, and it also extinguishes fire.

CHEMISTRY OF THE FURNACE.

CARBONIC OXIDE is a colorless, transparent, combustible gas, wdich burns with a pale blue flame, as may be seen at times on opening a locomotive fire-box door. Its presence in a furnace is evidence of imperfect combustion from a deficient supply of air, as it indicates that only eight parts of oxygen instead of sixteen parts have united with six parts of carbon.

TABLE.

The following table exhibits the comparative amounts of water which can be, under perfect conditions, evaporated from the substances named:

One pound burned.	Water evaporated.
Hydrogen	64.28
Carbon (average of several experiments)	14.77
Carbonic Oxide	4.48
Sulphur	4.18
Alcohol	13.40
Oil gas	22.11
Turpentine	20.26

The last four substances are compounds, and the last three consist almost wholly, or chiefly of carbon and hydrogen. The total heating power of average coal is, it may be noted to advantage, about 12.83 pounds of water upon the same conditions as above described. Hydrogen, it is seen, stands pre-eminently at the head of the list for heating power, represented by the evaporation of 64¼ pounds of water, whilst carbon, the next in order, and the staple combustible element of fuel, has only a heating power of 14¾ pounds of water.

HEAT-PROOF AND ORNAMENTAL PAINTS.

Steam pipes, boiler fronts, smoke connections and iron chimneys are often so highly heated that the paint upon them burns, changes color, blisters and often flakes off. After long protracted use under varying circumstances, it has been found that a silica-graphite paint is well adapted to overcome these evils. Nothing but *boiled linseed oil* is required to thin the paint to the desired consistency for application, no dryer being necessary. The paint is applied in the usual manner witn an ordinary brush. The color, of course, is black.

Another paint, which admits of some variety in color, is made by mixing soapstone, in a state of fine powder, with a quick-drying varnish of great tenacity and hardness. This will give the painted object a seemingly-enameled surface, which is durable and not affected by heat, acids, or the action of the atmosphere. When applied to wood it prevents rotting, and it arrests disintegration when applied to stone. It is well known that the inside of an iron ship is much more severely affected by corrosion than the outside, *and this paint has proven itself to be a most efficient protection from inside corrosion.* It is light, of fine grain, can be tinted with suitable pigments, spreads easily, and takes hold of the fibre of the iron or steel quickly and tenaciously.

Turpentine well mixed with black varnish also makes a good coating for iron smoke pipes.

Much brighter and more pleasant appearing engine rooms can be made by making the surfaces white. Lime is a good non-conductor of heat and it has the further quality of protecting iron from rust, so it would appear that whitewash was as good a material with which to cover boiler fronts, smoke stacks, steam pipes, etc., as any other substance.

To prepare whitewash for this purpose it is only necessary to add a little salt or glue to the water used for dissolving the lime, as either of these substances will make it stick readily and it cannot afterward be easily rubbed off; but perhaps the best way to prepare the whitewash would be to boil a pound of rice until it has become the consistency of starch, all of the solid particles having been broken up by boiling, and add this solution to the solution of lime in water.

This last preparation is also very good for outside work, for after it has been applied and has an opportunity to dry, no amount of rain will wash it off and its appearance is almost equal to white paint, and no amount of heat ordinarily met with will discolor it, although the heat of the fire box doors, if it was applied in such place, would give it a brownish cast of color. Even the brick setting of a boiler looks very much better when nicely whitewashed than when of its natural color, and if the ceiling and walls of the boiler room are also whitewashed the effect is quite pleasing, more healthful and conduces greatly to cleanliness.

Any engineer who tries this, renewing the whitewash as frequently as he would paint, will give this plan of painting pipes and boiler front the preference over the use of any kind of black paint.

PRESSURE RECORDING GAUGE.

Fig. 105.

This device is an ingenious mechanism actuated by clock work and the varying pressures of steam formed within the boiler; it records the time and the pressure upon a revolving roll of paper and preserves an accurate account of the varying conditions which have existed within the boiler.

The advantages derived from its use may be thus summarized: 1, It is a monitor constantly teaching the fireman to be careful to maintain an equal pressure of steam. 2, This uniform steam made possible by the use of the gauge is productive of the greatest possible economy.

PRESSURE RECORDING GUAGE.

3, The even strain maintained insures a long life to the boiler and a minimum of repairs. 4, It is the vindication of an attentive and careful fireman and allows him due credit for his skill and faithfulness, which is too often ill appreciated for lack of a reliable record.

Although described as a boiler room fixture, where it is frequently found in position, the proper place for this admirable device is in the steam user's office, thus establishing *a nerve connection,* between engineer and owner, relating to the safety and economy of the power-plant to their mutual great advantage.

HORSE POWER AS APPLIED TO BOILERS.

By general agreement a horse power as applied to steam boilers is thirty (30) pounds of feed water at a temperature of 100 degrees Fahr. converted into steam in 1 hour at 70 pounds gauge pressure.

The standard is all that can be asked because the same test will determine two things; first the steam making capacity of the boiler and second its evaporative efficiency, which is all that is necessary to know in determining the commercial rating of boilers.

But it is a fact that, without an engine attached, there is no such thing as calculating the horse power of a boiler upon general principles. A well constructed engine with a given pressure of steam upon a piston of a given area and moving at a certain velocity in feet per minute, will always and under all conditions develop the same power so long as the boiler is able to furnish a sufficient quantity of steam to keep up that pressure; and it matters not whether the steam is taken from a boiler rated at 60 horse power or 30.

An evidence of the fact that there is no standard rule for calculating the horse power of boilers that can be depended upon, is that no two engine builders send out the same sized boilers with the engine of the same rated power. Experience has taught them that to furnish steam sufficient to work their engines up to their ratings that a certain sized boiler is required, and what would be considered 30 horse power by one manufac-

HORSE POWER AS APPLIED TO BOILERS.

turer might be considered 35 or more by another—the difference being in the economy of the engine of using the steam, and not in the boiler for making it.

Then, again, a boiler that might furnish a sufficient quantity of steam to work a certain type of engine up to 40 horse power without forcing the fire might, with another style of engine, in order to generate the same power and perform the same duty, require to be forced beyond the limits of safety or economy. Therefore, considering the varying conditions under which all steam boilers are placed, there is no such a thing as any reliable standard rule for calculating the horse power of boilers, but only an approximate one at the best.

Hence it is best to select an engine of a certain power, and then let the same manufacturers furnish a boiler to correspond with it; and so long as the two are adapted to each other and the boiler of sufficient capacity to work the engine up to its full ratings, it matters but little whether the boiler figures the same horse power or not.

It has been found in practice that it is not good economy to carry pressure higher than eighty pounds in single cylinder automatic cut off engines.

As pressures increase, it becomes possible to use more economical engines, reducing water consumption per horse power per hour, thus requiring a smaller amount of heating surface and grate surface, that is to say, a smaller boiler and furnace for a given power.

For pressure between eighty and one hundred and twenty pounds, the compound engine gives the best results, while for higher pressures triple and quadruple expansion engines are the most economical.

Rule for Estimating Horse Power of Horizontal Tubular Steam Boilers.

Find the square feet of *heating surface* in the shell, heads and tubes, and divide by 15 for the nominal horse power.

The office of a boiler is to make steam and its real efficiency or the measure of its utility to the purchaser is measured

HORSE POWER AS APPLIED TO BOILERS.

by the amount of water it can turn into steam in a certain length of time and the amount of coal it requires to do this work.

An ordinary 54"x16' boiler with forty 4" tubes, 25 sq. ft. of grate surface and 800 sq. ft. of heating surface, in a general way is a 75 h. p. boiler, but good practice will get from it 100 h. p., and the very best modern engines 200 h. p.

BOILER SETTING.

The method, either ill or good in which steam boilers are "set" or arranged in their brick work and connections, will vary the quantity of fuel used by as much as one-fifth ; hence the importance of knowing the correct principles upon which the work should be done.

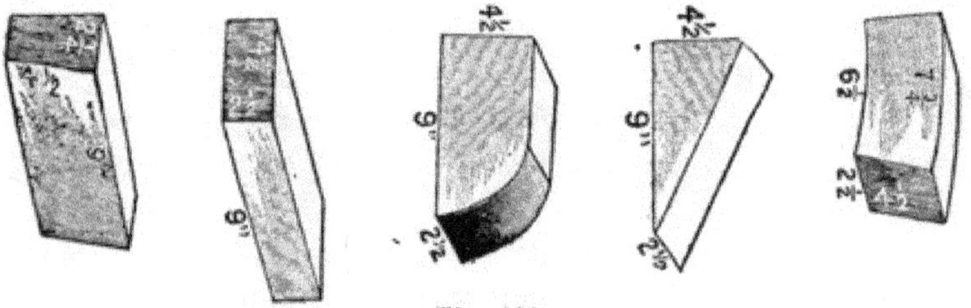

Fig. 106.

The portion of the steam plant called "the boiler" is composed of two parts—the boiler and *the furnace*, and the latter may be considered a part of the "setting" as it is mainly composed of brick work.

Two kinds of brick are used in boiler setting—the common brick for walls, foundations and backing to the furnace, and so-called fire-brick, which should be laid at every point where the fire operates directly upon the furnace and passages.

Fire brick should be used in all parts of the setting which are exposed to the hot gases. It is better to have fire brick lining tied in with red brickwork, unless the lining is made 13½ inches thick, when it can be built up separate from outside walls. This arrangement will require very heavy walls. As usual, but 9 inches fire brick lining is used in the fireplace

BOILER SETTING.

and 4½ inches behind the bridge wall. Joints in the fire brick-work should be as thin as possible.

Fig. 106 represents some of the different shapes in which fire brick are made to fit the side of the furnace. They are called by special names indicated by their peculiar form, circle-brick, angle-brick, jamb-brick, arch-brick, etc. The common fire brick are 9″x4½″x2½″ in size, as shown in the figure.

The peculiar quality in fire bricks is their power to resist for a long time the highest temperatures without fusion; they should be capable of being subjected to sudden changes of temperature without injury, and they should be able to resist the action of melted copper or iron slag. Fire brick are cemented together with fire clay which is quite unlike the ordinary mortar which is most suitable for common brick.

The setting as well as construction of boilers differs greatly, but in all the end to be sought for is *a high furnace heat,* with as little *waste as possible, at the chimney end.* To attain this there must be (1) a sufficient thickness of wall around the furnace, including the bridge, to retain as nearly as may be every unit of heat. (2) A due mixture of air admitted at the proper time and temperature to the furnace. (3) A proportionate area between the boiler and the surface of the grates for the proper mixing of the gases arising from combustion. (4) A correct proportion between the grate surface, the total area of the tubes and the height and area of the chimney.

The principal parts and appendages of a furnace are as follows:

The furnace proper or fire box, being the chamber in which the solid constituents of the fuel and the whole or part of its gaseous constituents are consumed.

The grate, which is composed of alternate bars and spaces, to support the fuel and to admit the air.

The dead plate, that part of the bottom of the furnace which consists of an iron plate simply.

The mouth piece, through which the fuel is introduced and often some air. The lower side of the mouth piece is the dead plate.

BOILER SETTING.

The fire door: Sometimes the duty of the fire door is performed by a heap of fuel closing up the mouth of the furnace.

The furnace front is above and on either side of the fire door.

The ash pit. As a general rule the ash pit is level, or nearly so, with the floor on which the fireman stands, and as for convenient firing, the grate should not be higher than 28 to 30 inches, the depth of ash pit is thereby determined.

The ash pit door is used to regulate the admission of air.

The bridge wall.

The combustion or flame chamber.

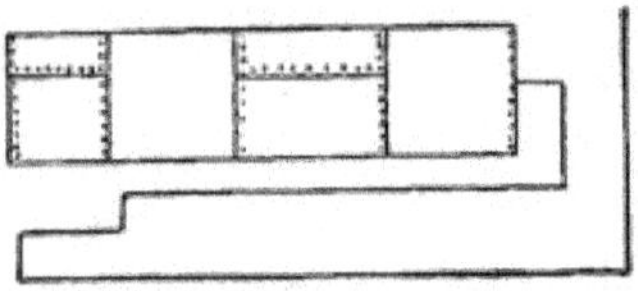

Fig. 107.

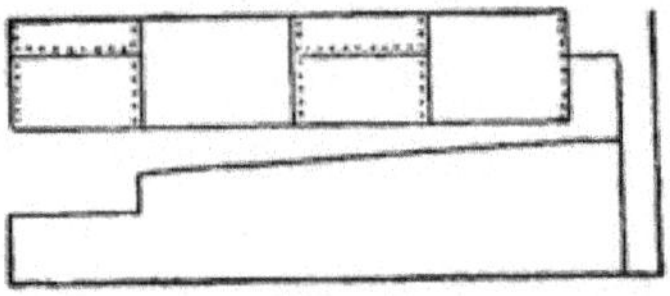

Fig. 109.

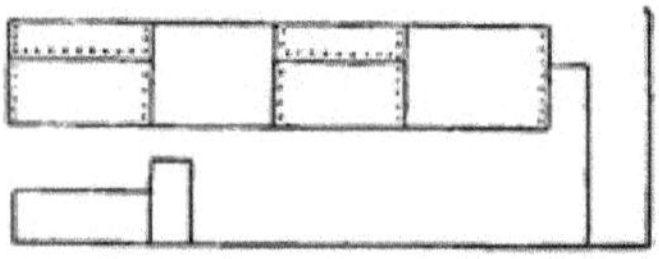

Fig. 108.

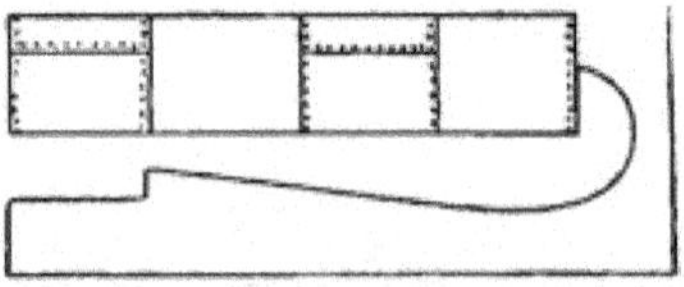

Fig. 110.

The arrangement of the space behind the bridge wall is found usually to be in some one of the following forms: Level from bridge wall to back (Fig. 107). A square box, depth ranging from 15 inches to 6 feet (Fig. 108). A gradual rise from bridge to back end of boiler, where only six inches is found and generally circular in form (Fig. 109). A gradual slope toward back, leaving a distance of about 36 inches from boiler (Fig. 110).

The advocates of Fig. 107 claim that the office of the flame is to get into as close contact with the bottom as possible, and this form compels the flame to do so. In burning soft coal this form is found to soot up the bottom of the boiler very badly.

BOILER SETTING.

Fig. 108 is followed more extensively than any other, the variations being the depth of chamber; with depth generally from 36 to 40 inches.

Fig. 109 has nothing to commend it, except in cases where bridge is too low.

Fig. 110 is followed a great deal and gives very good satisfaction. This form allows for the theory of combustion, namely, the expansion of the gases after leaving bridge wall.

Space behind the bridge wall should be enlarged, as it will reduce the velocity of fire gases, and thus have them give up more of their heat to the boiler.

The bridge wall should not be less than 18 inches at bottom, but may be tapered off toward top to 9 or 13 inches.

SETTING OF WATER TUBE BOILERS.

On page 67, Fig. 26, is exhibited a steam boiler with inclined tubes. The setting is this style of boilers is as follows:

A brick wall in laid for the front with suitable openings for the doors of the furnace and ash pit, and protected on the outside by a front of cast iron, and on the inside by a lining of fire brick.

At the back of the grates a bridge wall is run up to the bottom of the inclined water tubes, so that the hot gases that arise over it must circulate among the tubes.

A counter wall is laid on an incline from the top of the tubes to the back of the drum. This is laid on perforated plates or bars and is covered with fire brick. A wall is also built at the lower and back end of the tubes to carry them.

Back of the whole is the outer wall with openings for giving access to the tubes and smoke chambers. Side walls are raised to enclose the same and are arched at the top to come nearly in contact with the drum, which is carried partly by brackets and partly by the connections to the tubes.

POINTS RELATING TO BOILER SETTING.

Long and heavy boilers are best suspended from two beams or girders by two or three bolts at each end. Boilers over 40 feet long should have three or even four sets of hangers, as the case may require.

BOILER SETTING.

Side brackets resting on masonry may be used for short boilers. If used on long boilers, side plates or expansion rollers should be used at one end of boiler. There ought to be not more than two brackets on one side, so divided that the distance between them is about three-fifths of the total length of the boiler, or the distance from ends of boiler to center of bracket is equal to one-fifth the length of boiler.

The side walls in boiler-setting should not be less than twenty inches with a two inch air space; the rear wall may vary from 12 to 16 inches according to the size of the boiler; the front wall 9 inches and the bridge wall may be from 18 to 24 and perfectly straight across the rear of the furnace. If the boilers are supported by side walls, the outside walls should be not less than 13 inches thick and have pilasters where the boiler is resting.

Flues touching the boiler above the water space should be emphatically condemned.

Unless the boiler walls are very heavy, they should be stayed by cast or wrought iron bunch stays, held together by rods at tops and bottoms.

It is dangerous to have large spaces in which gases may collect for sudden ignition, producing the so-called "back draft."

Connections between the rear end of the boiler and brickwork is best made with cast-iron plates or fire-brick, suspended, when boilers are suspended, as the expansion and contraction will destroy an arch in a short time. If resting on mud-drum stand, this connection can be arched, as in this case the rear end of boiler will remain stationary.

If the drafts from the different boilers come in the same direction, or nearly so, no special provision is necessary, but if the draft enters from directly opposite directions a center wall should be provided.

An advantage claimed for water in the ash pit is: by the dropping of hot ashes and cinders from the grate into the water.

BOILER SETTING.

steam is generated, which, in passing through the hot coal lying on the grate, is there divided into oxygen and hydrogen, thus helping the combustion.

A dry brick will absorb a pound of water, and it is the water in the mortar that causes it to set, and harden. To prevent this loss of the water of crystalization, and give it time to harden and adhere to the brick, the brick must be well saturated with water, before they are laid.

Whenever steam is allowed to come in contact with mortar or cement an injurious effect is produced. The action of the steam is much more rapid than that of air and water, or water alone, when in abundance, as the effect of the steam in every case is to soften the mortar and penetrate to a greater depth than water could possibly do.

The distance between the rear head of the boiler and brick-work should not be less than 12 inches.

In setting steam boilers, allowance must be made for the expansion and contraction of the structure and this is usually done by placing rollers under the rear lug or side bearing of the boiler. Care should be exercised that the boiler rests are always in good condition so that they may move freely and not place the boiler in any danger of sticking and buckling.

Kindling a Furnace Fire.

In kindling a coal fire in a furnace the phosphorus of a match inflames at so low a temperature (150 degrees Fahr.) that mere friction ignites it, and in burning (combining with oxygen of the air) it gives out heat enough to raise the sulphur of the match to the temperature of ignition (500 degrees Fahr.), which, combining in its turn with the oxygen of the atmosphere, gives out sufficient heat to raise the temperature of the wood to the point of ignition (800 degrees Fahr.), and at this temperature the wood combines with oxygen supplied by the air, giving out a temperature sufficient to raise the coal to the

KINDLING A FURNACE FIRE.

point of ignition (1000 degrees Fahr.), and the coal then combines with the free oxygen of the air, the ensuing temperature in the furnace varying, according to circumstances, from 3000 degrees to 4000 degrees Fahr. Thus we see that the ignition of the coal is the last of a series of progressive steps, each increasing in temperature.

And in each step it will be noted that a combination of oxygen is the essential connecting link and *that the oxygen is supplied in each instance at the same average temperature*—this fact contains a "point" relating to supplying furnaces with so called "hot air."

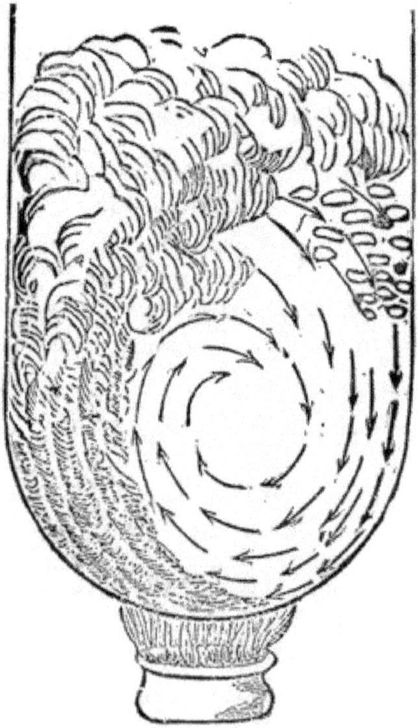

GAS PIPE.

Fig. 111.

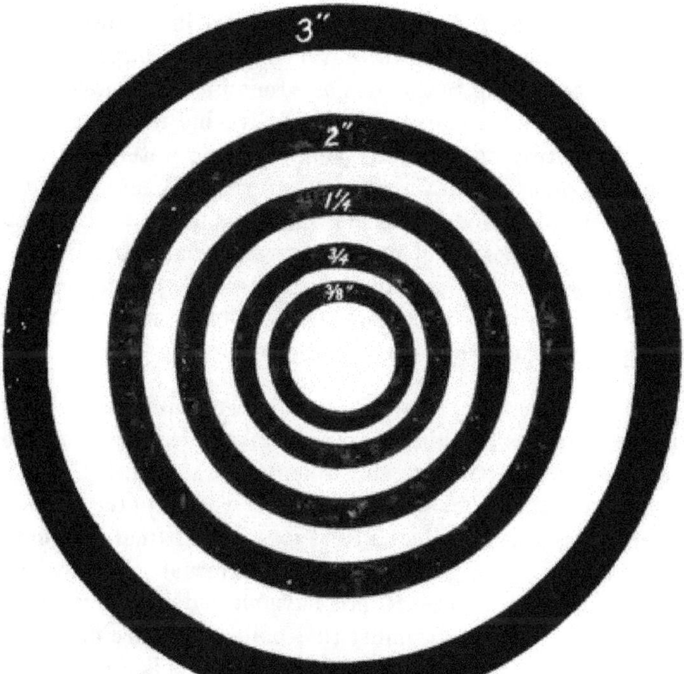

Fig. 112.

PIPES AND PIPING.

Next in importance after the skill necessary for the steam generator and the engine, is the proper arrangement and care and management of the pipes and valves belonging to a steam plant.

It is the first thing an engineer does in taking charge of a new place, to ascertain the exact course and operation of the water, steam, drain and other pipes.

Examiners for licensing marine and land engineers base their questions much more to ascertain the applicant's knowledge of piping than is generally known; hence the importance of the "points" in the succeeding pages relating to this subject.

Pipes are used for very many purposes in connection with the boiler room, and of course vary in size, in material and in strength, according to the purposes for which they are designed. There are pipes for conveying and delivering illuminating gas; pipes for conveying and delivering drinking water, and for fire purposes; pipes for draining and carrying off sewage and surface water; pipes for delivering hot water under high pressure, for heating purposes and power; pipes for delivering live steam under pressure, for heating purposes and power; pipes for delivering compressed air, for purposes of power and ventilation; pipes for conveying mineral oils, etc.

In Figs. 111, 112 113 and 114 are given approximate sizes of gas pipe and boiler tubes, taken from the catalogue of one of the oldest steamfitting establishments in the country. It will be observed that the size of gas pipe is computed from the internal diameter, while boiler tubes are estimated from the outside: thus, 3 in. gas pipe has an external diameter of 3½ inches, while 3 in. boiler tubes have an outside diameter of 3 inches only. It may be noted that boiler-tubes are made much more accurately as to size than gas pipe; this is especially true of the outside surfaces which are much smoother in one case than in the other.

BOILER TUBES.

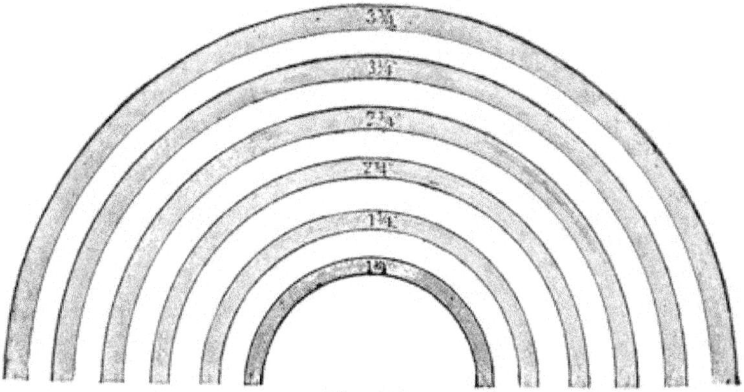

Fig. 113.

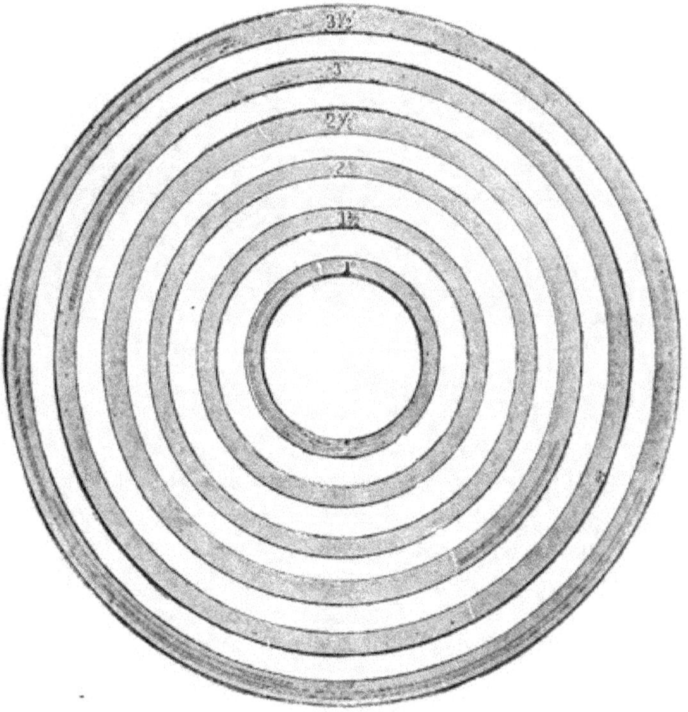

Fig. 114.

SURFACES AND CAPACITIES OF PIPES.

SIZES OF PIPES.	½ in.	¾ in.	1 in.	1¼ in.	1½ in.	2 in.	2½ in.	3 in.	3½ in.	4 in.	4½ in.	5 in.
1. Outside circumferences of pipes in inches..	2.652	3.299	4.136	5.215	5.969	7.461	9.032	10 99	12.56	14.13	15.70	17 47
2. Length of Pipe in feet to give a square foot of outside surface..	4.52	3.63	2.90	2.30	2.01	1.61	1.32	1.09	.954	.849	7.63	.686
3. Number of square feet of outside surface in ten lineal feet of Pipe..........	2.21	2.74	3.44	4.34	4.97	6.21	7.52	9.16	10.44	11.78	13.09	16.53
4. Cubic in. of internal capacity in ten lineal feet of pipe.....	36.5	63.9	103.5	179.5	244.5	402.6	573.9	886.6	1186.4	1527.6	1912.6	2398.8
5. Weight in lbs. of water in ten lineal feet of pipe.............	1.38	2.31	3.75	6.5	8.8	14.6	20.8	32 1	43.6	55.4	69 3	86 9

Pipe manufactured from double thick iro.ı is called X-strong pipe, and pipe made double the thickness of X-strong is known as XX-strong pipe. Both X-strong and XX-strong pipe are furnished plain ends—no threads, unless specially ordered.

The table "Data relating to iron pipe" will be found especially useful to the engineer and steam fitter. The size of pipes referred to in the table range from $\frac{1}{8}$ to 10 inches in diameter. In the successive columns are given the figures for the following important information:

1. Inside diameter of each size.
2. Outside diameter of each size.
3. External circumference of each size.
4. Length of pipe per square foot of outside surface.
5. Internal area of each size.
6. External area of each size.
7. Length of pipe containing one cubic foot.
8. Weight per foot of length of pipes.
9. Number of threads per inch of screw.
10. Contents in gallons (U. S. measure) per foot.
11. Weight of water per foot of length.

DATA

Relating to Iron Pipe.

Inside Diameter.	Outside Diameter.	External Circumference.	Length of Pipe per sq. ft. of Outside Surface.	Internal Area.	External Area.	Length of Pipe containing one Cubic Foot.	Weight per ft. of Length.	No. of Threads per inch of Screw	Contents in *Gallons per foot.	Weight of Water per foot of Length.
Inches.	Inches.	Inches.	Feet.	Inches.	Inches.	Feet.	Lbs.			Lbs.
1/8	.40	1.272	9.44	.012	.129	2500.	.24	27	.0006	.005
1/4	[illegible]	1.696	7.075	.049	.229	1385.	.42	18	.0026	.021
3/8	.67	2.121	5.657	.110	.358	751.5	.56	18	.0057	.047
1/2	.84	2.652	4.502	.196	.554	472.4	.84	14	.0102	.085
3/4	1.05	3.299	3.637	.441	.866	270.	1.12	14	.0230	.190
1	1.31	4.134	2.903	.785	1.357	166.9	1.67	11½	.0408	.349
1¼	1.66	5.215	2.301	1.227	2.164	96.25	2.25	11½	.0638	.527
1½	1.9	5.969	2.01	1.767	2.835	70.65	2.69	11½	.0918	.760
2	2.37	7.461	1.611	3.141	4.430	42.36	3.66	11½	.1632	1.356
2½	2.87	9.032	1.328	4.908	6.491	30.11	5.77	8	.2550	2.116
3	3.5	10.996	1.091	7.068	9.621	19.49	7.54	8	.3673	3.049
3½	4.	12.566	.955	9.621	12.566	14.56	9.05	8	.4998	4.155
4	4.5	14.137	.849	12.566	15.904	11.31	10.72	8	.6528	5.405
4½	5.	15.708	.765	15.904	19.635	9.03	12.49	8	.8263	6.851
5	5.56	17.475	.629	19.635	24.299	7.20	14.56	8	1.020	8.500
6	6.62	20.813	.577	28.274	34.471	4.98	18.76	8	1.469	12.312
7	7.62	23.954	.505	38.484	45.663	3.72	23.41	8	1.999	16.662
8	8.62	27.096	.444	50.265	58.426	2.88	28.34	8	2.611	21.750
9	9.68	30.443	.394	63.617	73.715	2.26	34.67	8	3.300	27.500
10	10.75	33.000	.355	78.540	90.792	1.80	40.64	8	4.081	34.000

* The Standard U. S. gallon of 231 cubic inches.

PIPES AND PIPING.

The division of process in the manufacture of pipe, takes place at $1\frac{1}{4}$ inch, $1\frac{1}{4}$ inch and smaller sizes being called butt-welded pipe, and $1\frac{1}{2}$ inch and larger sizes being known as lap-welded pipe; this rule holds good for standard, X-strong and XX-strong.

JOINTS OF PIPES AND FITTINGS.

The accompanying illustrations represent certain joints, couplings and connections used in steam and hot water heating systems.

For many years in the matter of pipe joints there has been little change. The cast-iron hub and spigot joint, Fig. 115, caulked with iron borings, is probably the oldest kind of joint. This is still generally adopted in hot water heating of a certain class, and was formerly used with low-pressure steam. A fairly regular smooth internal service is obtained, and once made tight is very durable. Cast-iron flanged pipes have also been a long time in use. These joints are made with a wrought-iron ring gasket, wrapped closely with yarn, Fig. 116, which is sometimes dipped in a mixture of red and white lead. It is placed between the flanges, it being of such a diameter as to fit within the bolts by which the joint was screwed up and a nest or iron joint, B B, caulked outside the annular gasket between the faces of the flanges.

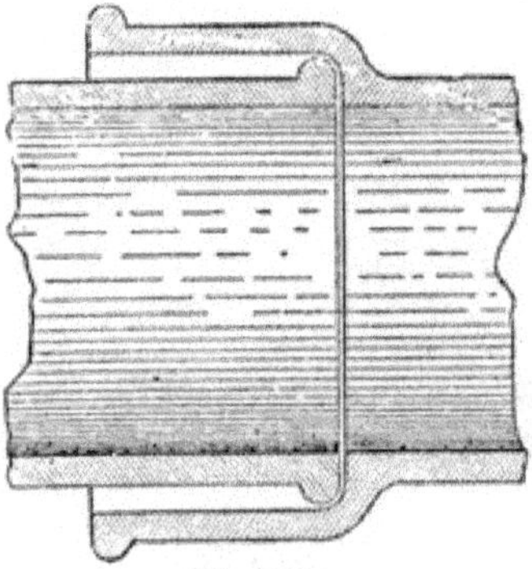

Fig. 115.

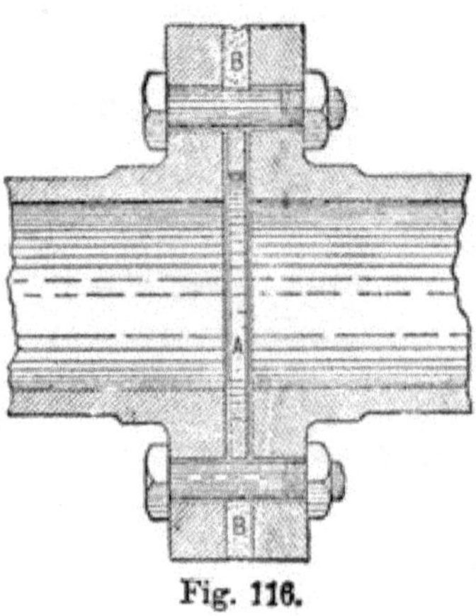

Fig. 116.

The next step in cast-iron flange pipe joints was the facing or turning up of the flanges and the use of a gasket of rubber, copper, paper or cement, with

PIPES AND PIPING.

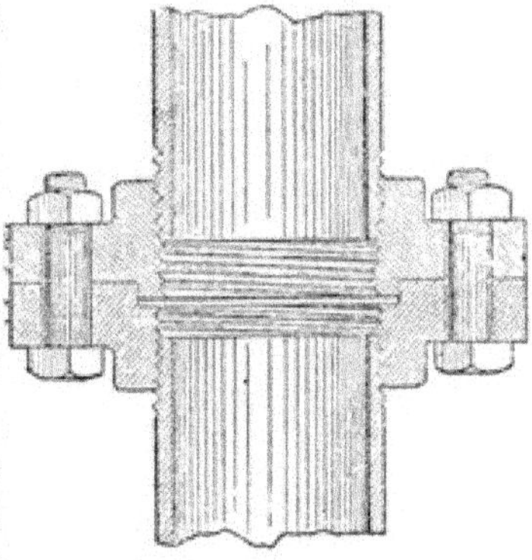

Fig. 117.

bolts for drawing the faces together. These joints for cast iron pipes have not been changed excepting for some classes of work where a lip and recess, Fig. 117, formed on opposite flanges, which makes the internal surfaces smooth and aid in preventing the gaskets from being blown out.

The introduction of wrought iron welded pipes has diminished the use of cast-iron pipes for many purposes, especially in heating apparatus and other pipe systems. Its advantages are lightness, the ease with which various lengths can be obtained and its strength. In wrought-iron pipe work the general practice in making joints between pipes is a wrought-iron coupling, Fig. 118, with tapered threads at both ends. The pipes do not meet at their ends, and a recess of about ⅜ inch or more long by the depth of the thickness of the pipes is left at every pipe end. A similar tapered thread is used in connecting the cast-iron fittings, elbows, tees, etc., Fig. 119, to the pipe, and a large recess is necessary in each fitting to allow for the tapping of the threads. Thus the inside diameter of the fitting is larger by ⅛ inch than the outside diameter of the pipe, and the internal projection of the thickness of the pipe and that of the thread of the fitting increases materially the friction due to the interior surfaces of pipe and

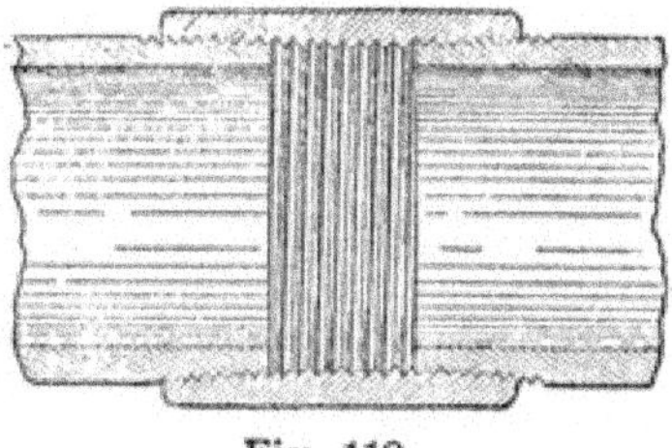

Fig. 118.

Fig. 119.

Fig. 120.

PIPES AND PIPING.

fitting. This class of joint requires care in the tapping of the fittings and in the cutting of tapered threads on the pipes; much trouble is caused by an inaccurately cut thread, as it may throw a line of pipes several inches out of place and put fittings and joints under undue and irregular strains.

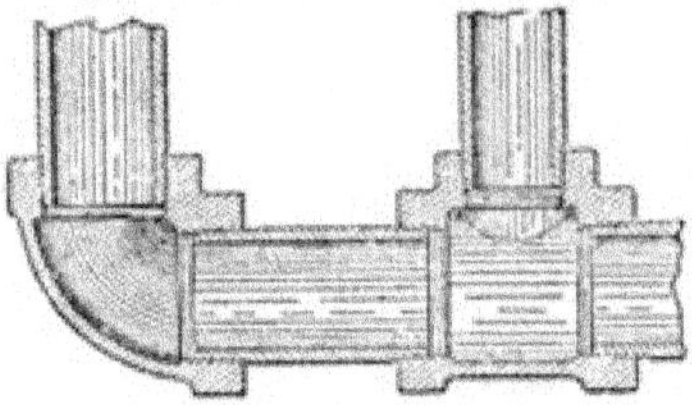

Fig. 121.

The right and left threaded nipple, Fig. 119, is used as a finishing connection joint and between fittings. Space equal to the length of the two threads is required between the two fittings to be connected in order to enter the nipple, and one or both fittings should be free to move in a straight line when the nipple is being screwed up. To make up this joint time and care are necessary. The right threaded end on nipple should be first firmly screwed with the tongs or wrench into the right threaded end of fitting, then slacked out and screwed up again by hand until tight, when it is screwed back by hand, at the same time counting the number of threads it has entered by hand. The same is done with the left threaded end of nipple and fitting. If the right and left threads of nipple have counted the same number of threads, each thread, when making the joint up, should enter the fittings at the same time if possible, and particular care must be taken that the fittings are exactly opposite, to facilitate catching on, prevent crossing threads, and that no irregular strain comes on the nipple while being screwed up.

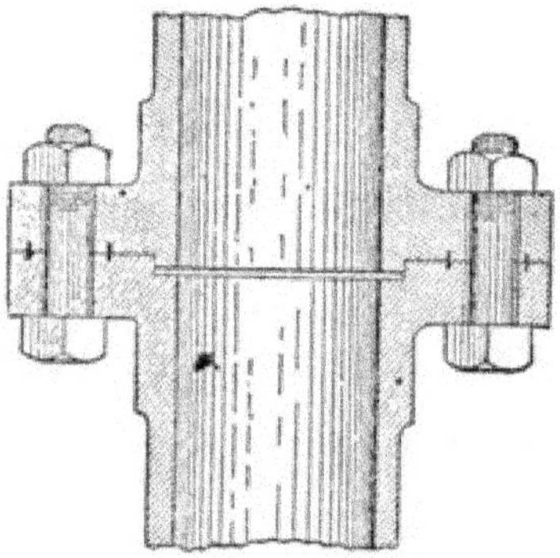

Fig. 122.

In screwing up these nipples the coupling has to be turned with flats on the external surface to fit an internal wrench: in such cases the thread on nipple has one continuous taper.

PIPES AND PIPING.

These special couplings are marked with ribs on the outside to distinguish them. Fig. 120 represents another joint in wrought-iron piping known as the "union," composed of three pieces and the washer. Unions are also made with ground joints, and the washer dispensed with. Radiator valves are now generally connected by them, but if the hole in the radiator is not tapped accurately, the union when drawn up will not be tight, or if tight, the valve will not be straight.

Fig. 121 shows right and left threaded nipple connecting elbow and tee with wrought-iron pipes.

The flange union, Fig. 122, is another joint generally used on wrought-iron pipes above 4 or 5 inches in diameter in making connections to valves, etc., and on smaller pipes in positions where it is a convenient joint. This joint consists of two circular cast-iron flanges with the requisite number of holes for bolts, and central hole tapped tapered to receive thread of pipe. The abutting faces of the flanges are generally turned and the holding bolts fitted into the holes.

STEAM AND HOT WATER HEATING.

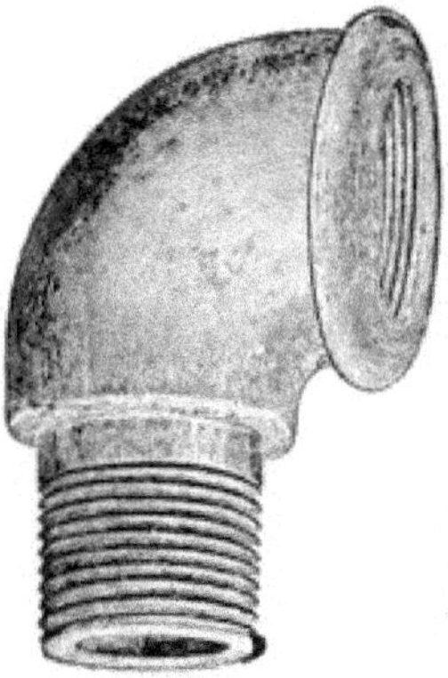

Fig. 123.

Fig. 124.

The heating by means of pipes through which are conveyed hot water and steam is a science by itself and yet one claiming some degree of familiarity by all engineers, steam users, and architects.

In practice it requires a knowledge of steam, air and temperatures, of pressure and supply; a familiarity with heat and heating surfaces and with all contrivances, appliances and devices that enter into the

STEAM AND HOT WATER HEATING.

warming and ventilation of buildings. So long as factories, public and private buildings are erected, so long will warming and ventilation keep progress with steam engineering and remain a part of the general mechanical science required of the supervisory and practical engineer.

In what is called *the system of open circulation,* a supply main conveys the steam to the radiating surfaces, whence *a return main conducts the condensed water either into an open tank for feeding the boiler, or into a drain to run to waste,* the boiler being fed from some other source; the system of what is called *closed circulation* is carried out either with separate supply and return mains, both of which extend to the furthest distance to which the heat has to be distributed, or else with a single main, which answers at once for both the supply and the return, either with or without a longitudinal partition inside it for separating the outward current of steam supply from the return current of condensed water.

In either case suitable traps have to be provided on the return main, *for preserving the steam pressure within the supply main and radiators.* These two systems, in any of their modifications, may also be combined, as is most generally done in any extensive warming apparatus.

The system of closed circulation requires the boiler to be placed so low as will allow all the return pipes to drain freely back to it above its water-level. This condition has been modified mechanically by the automatic "trap," a device frequently employed for lifting from a lower level, part or all of the condensed water, and delivering it into the boiler; it is, in fact, a displacement pump.

The same result has been attained by draining into a closed tank, placed low enough to accommodate all the return pipes, and made strong enough to stand the full boiler pressure with safety, and then employing a steam pump, either reciprocating or centriful, to raise the water from this tank to the proper level for enabling it to flow back into the boiler, the whole of the circulation being closed from communication with the atmosphere.

STEAM AND HOT WATER HEATING.

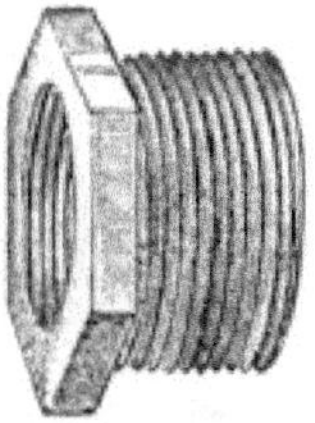

Fig. 125.

Fig. 126.

Fig. 127.

There are two systems of steam heating, known as the *direct* and the *indirect* system.

Direct radiating surfaces embrace all heaters placed within a room or building to warm the air, and are not directly connected with a system of ventilation.

Indirect radiation embraces all heating surfaces placed outside the rooms to be heated, and can only be used in connection with some system of ventilation.

For warming by direct radiation, the radiators usually consist of coils, composed of $\frac{3}{4}$-inch and 1-inch steam pipes, which are arranged in parallel lines and are coupled to branch tees or heads. In a few exceptional cases, radiators of peculiar shapes are specially constructed. In all cases the coils must have either vertical or horizontal elbows of moderate length, for allowing each pipe to expand separately and freely. Sometimes short lengths of pipe are coupled by return-bends, doubling backwards and forwards in several replications one above another, and forming what are called "return-bend coils," and when several of these sections are connected by branch tees into a compact mass of tubing, the whole is known as a "box-coil."

Steam and Hot Water heating have long been acknowledged as altogether most practical and economical in every way—and their universal adoption in all the better class of buildings throughout the country is positive proof of their superiority.

STEAM AND HOT WATER HEATING.

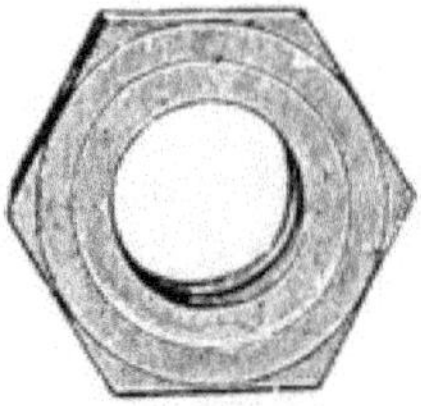

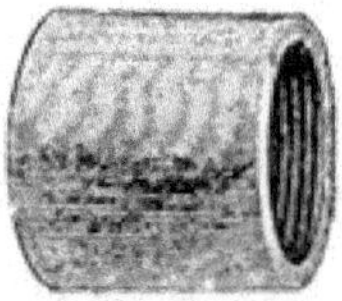

Fig. 128. Fig. 129. Fig. 130.

The heat from steam is almost exactly indentical with that from hot water, and few can distinguish between the two systems when properly erected.

They are both healthful, economical and satisfactory methods of warming. They give no gas, dust nor smoke; are automatically regulated, and therefore allow of an even and constant temperature throughout the house, whatever be the condition of the weather outside.

The circulation of the steam through the warming pipes is effected in an almost unlimited variety of ways, and the cause producing the circulation throughout the pipes of the warming apparatus is solely the difference of pressure which results from the more or less rapid condensation of the steam in contact with the radiating surfaces.

A partial vacuum is formed by this difference of pressure *within the radiating portions of the apparatus*, and the column of steam or of water equivalent to this diminution of pressure, constitutes the effective head producing the flow of steam from the boiler, at the same time the return current of condensed water is determined by the downward inclination of the pipes for the return course.

Points Relating to Steam Heating.

No two pipes should discharge into a T from opposite directions, thus retarding the motion of both or one of the returning currents. This is called "butting" and is one of the most vexatious things to encounter in pipe fitting.

POINTS RELATING TO STEAM HEATING.

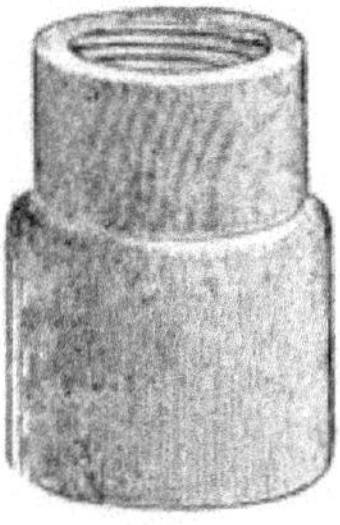

Fig. 131.

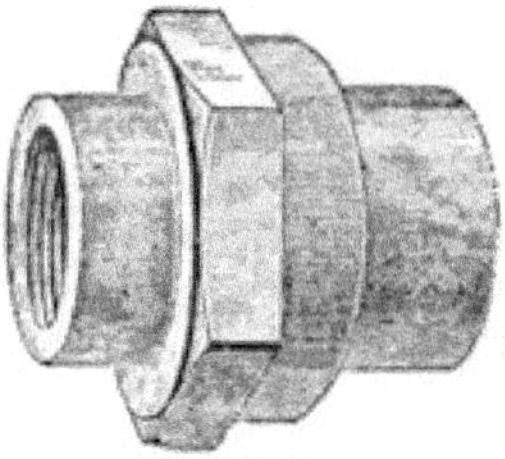

Fig. 132.

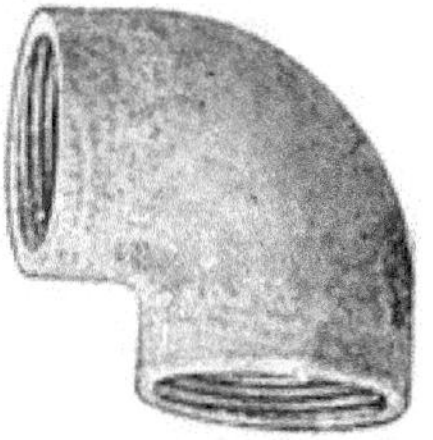

Fig. 133.

All steam piped rooms should be frequently dusted, cleaned and kept free from accumulation of inflammable material.

The use of the air valve is as follows: In generating steam from cold water all the free air is liberated and driven off into the pipe, with the air left in them, all of which is forced up to the highest point of the coils or radiators, and compressed equal to the steam pressure following it. Now, by placing a valve or vent at the return end of the pieces to be heated, the air will be driven out by the compression. Why the vent is placed at the return is, that the momentum of the steam, it being the lightest body, will pass in the direction of it, falling down into the return as it condenses, thus liberating the air. Otherwise, should the vent not work, and the air is left in the radiator, it will act as an air spring, and the contents of the pipes left stationary will be the result; no circulation, no heat; and the greater steam pressure put on, the greater the chances are of not getting any heat; and thus a little device, with an opening no larger than a fine needle, will start what a ton of pressure would not do in its absence.

If the drip and supply pipes are large there is very little danger of freezing, provided suitable precautions are taken to leave the pipes clear. They should be blown through, when left, and the steam valve should be closed. There should also be a free chance for air to escape in all systems of piping.

No rule can be given relating to capacity for heating pipes and radiators which do not require to be largely modified by surroundings.

POINTS RELATING TO STEAM HEATING.

The field of steam-heating would seem to be limitless—in one public building it required recently 480,000 dollars to meet the expenditures in this single line. As an example of warming on an extensive scale may be taken a large office in New York, of which the following are the particulars:

Total number of rooms, including halls and vaults.		286
Total area of floor surface....................	sq. ft.	137,370
Total volume of rooms.....................	cub. ft.	1,923,590

A second example is furnished by the State Lunatic Asylum at Indianapolis:

Length of frontage of building, more than.			2,000 lin. ft.
Total volume of rooms....................			2,574,084 cub. ft.
Warming Apparatus	indirect radiating surface...........	23,296	
	Direct..................	10,804	
	Total.........................		34,100 sq. ft.
Boilers,...	Grate area................. ..		180 sq. ft.
	Heating surface.............		5,863 sq. ft.

The "overhead" system of heating with steam pipes has several advantages. 1. The pipes are entirely out of the way. 2. They do not become covered with odds and ends of unused materials. 3. If they leak the drip fixes the exact location of place needed to be repaired. 4. The room occupied overhead cannot be well otherwise utilized, hence in shops the system has proved efficient.

But for offices or store rooms the overhead system is not approved of owing to the heat beating down upon the occupants and causing headache.

When overhead heating pipes are used, they should not be hung too near the ceiling. If the room be a high one, it is better to hang them below, rather than above, the level of the belts running across the room, and they should not be less than three or four feet from the wall.

STEAM HEATING.

It is important to protect all woodwork or other inflammable material around steam pipes from immediate contact with them, especially where pipes pass through floors and partitions. A metal thimble should be placed around the steam pipe, and firmly fastened on both sides of the floor, in such a way as to leave an air space around the steam pipe.

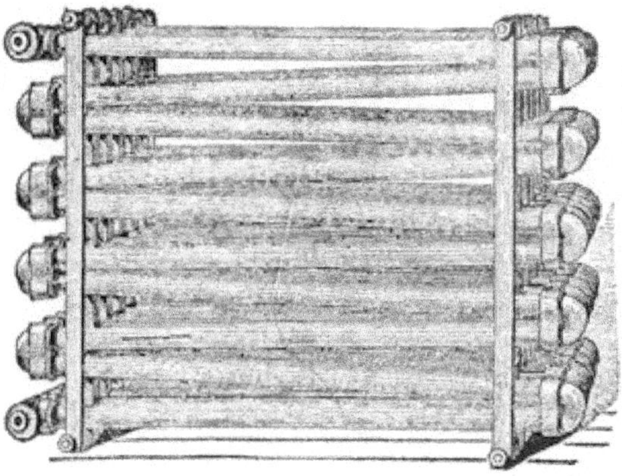

Fig. 134.

For indirect radiating surfaces, the box coils are the forms most used. The chambers or casings for containing them are made either of brickwork, or often of galvanized sheet-iron of No. 26 gauge, with folded joints. The coils are suspended freely within the chambers, which are themselves attached to the walls containing the air inlet flues. Besides coils of wrought iron tubes, cast-iron tablets or hollow slabs, having vertical surfaces with projecting studs or ribs, have been extensively used for the radiating surfaces.

As the amount of heat given off from the radiator cannot be satisfactorily controlled by throttling the steam supply, it is usual to divide all radiators into sections, each of which can be shut off from the supply and return mains, separately from the rest of the sections. This method of regulation applies to radiators for indirect heating as well as for direct.

Vertical pipe coils, constitute a distinctive form of radiator now largely used. In these a number of short upright 1-inch tubes, from 2 feet 8 inches to 2 feet 10 inches long, are screwed into a hollow cast-iron base or box; and are either connected together in pairs by return-bends at their upper ends, or else each tube stands singly with its upper end closed, and having a hoop iron partition extending up inside it from the bottom to nearly the top. The supply of steam is admitted into the bottom casting; and the steam on entering, being lighter than the air, ascends through one leg of each siphon pipe and descends through the other, while the condensed water trickles down either leg, and with it the displaced air sinks also into the bottom box. For getting rid of the air, a trap is provided, having an outlet controlled by metallic rods; as soon as all the air has escaped and the rods become heated by the presence of unmixed steam, their expansion closes the outlet.

A thorough drainage of steam pipes will effectually prevent cracking and pounding noises.

The windward side of buildings require more radiating surface than does the sheltered side.

When floor radiators are used, their location should be determined by circumstances; the best situations are usually near the walls of the room, in front of the windows. The cold air, which always creates an indraft around the window frames, is thus, to some extent, warmed as it passes over the radiators, and also assists in the general circulation.

Water of condensation will freeze quicker than water that has not been evaporated, for the reason that it has parted with all its air and is therefore solid.

Whatever the size of the circulating pipes, the supply and drip pipes should be large, to insure good circulation; the drip pipes especially so. This is all the more necessary when the pipes are exposed, or when there is danger of freezing after the steam is shut off.

It is important to see that no blisters or ragged pipes go into the returns, and also to make sure that the ends are not "blurred in" with a dull pipe cutter wheel so as to form a place of lodgment for loose matter in the pipe to stop against.

POINTS RELATING TO STEAM HEATING.

Fig. 135.

Fig. 136.

Fig. 137.

Experiments recently made on the strength of bent pipes have developed some things not commonly known, or at least not recognized, that is, the strain on the inside of the angles, *due to the effort of the pipes to straighten themselves under pressure.* The problem is one of considerable intricacy, resolvable, however, by computation, and is a good one for practice. In the experiment referred to, a copper pipe of $6\frac{3}{4}$ in. bore, $\frac{3}{16}$ in. thick, was used. The angle was 90 degrees, and the legs about 16 in. long from the center. At a pressure of 912 pounds to an inch, the deflection of the pipe was nearly $\frac{3}{8}$ in., showing an enormous strain on the inner side, in addition to the pressure.

Steam valves should be connected in such a manner that the valve closes against the constant steam pressure.

Interesting experiments show that the loss by condensation in carrying steam one mile is 5 per cent. of the capacity of the main, and a steam pressure of seventy-five pounds carried in five miles of mains, ending at a point one-half mile from the boiler house only shows a loss of pressure of two pounds.

In steam warming it is necessary to bring the water to a boiling point to get any heat whatever: in hot water warming, a low temperature will radiate a corresponding amount of heat.

POINTS RELATING TO STEAM HEATING.

Never use a valve in putting in a low pressure apparatus if it is possible to get along without it. All the valves or cocks that are actually required in a well-proportioned low pressure apparatus are, a cock to blow off the water and clean out the return pipes, another to turn on the feed water. Of course the safety valves, guage cocks, and those to shut fire regulators and such as are a part of the boiler, are not included in this "point."

The most important thing in connecting the relief to return pipes is, that it should always be carried down below the line, the same as all vertical return pipes. In connecting the reliefs, so that the lower opening can at any time be exposed to the steam, there will be the difficulty of having the steam going in one direction, and the water in another.

The relief pipe should "tap" the steam at its lowest or most depressed points. It should always be put in at the base of all steam "risers" taking steam to upper floors.

In leaving the boiler with main steam pipe, raise to a height that will allow of one inch fall from the boiler to every ten feet of running steam pipe; this is sufficient, and a greater fall or pitch will cause the condensed water in the pipe to make at times a disagreeable noise or "gurgling."

The flow pipe should never start from the boiler in a horizontal direction, as this will cause delay and trouble in the circulation. This pipe should always start in a vertical direction, even if it has to proceed horizontally within a short distance from the boiler. Reflection will show that the perfect apparatus is one that carries the flow pipe in a direct vertical line to the cylinder or tank; this is never, or but rarely possible, but skill and ingenuity should be exercised to carry the pipes as nearly as possible in this direction.

The flow of steam ought not to be fast enough to prevent the water of condensation from returning freely. All the circulating pipes should be lowest at the discharge end, and the inclination given them should not be less than one foot in fifty.

POINTS RELATING TO STEAM HEATING.

Fig. 138.

Fig. 139.

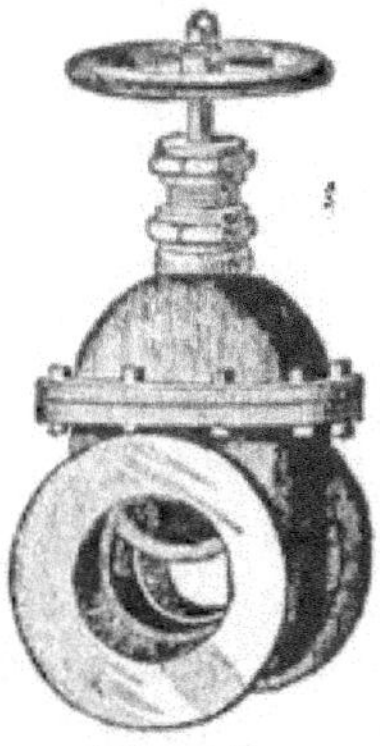
Fig. 140.

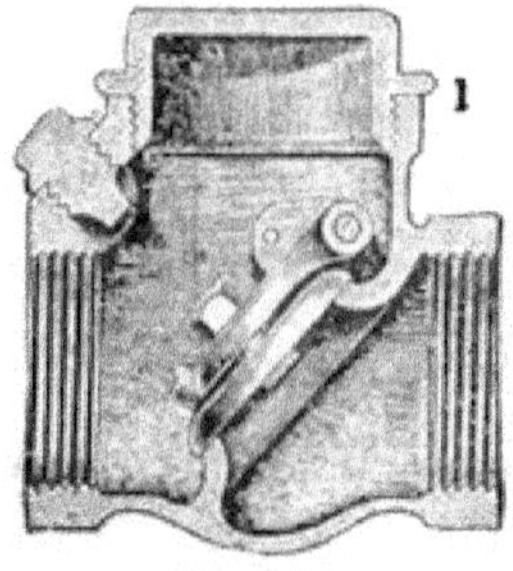

Fig. 141.

The general rule is to lay the main pipes from the boiler so that the pipe will drain from the boiler. Where this is done it is necessary to have a drip just before the steam enters the circulation. This drip is connected to a trap, or, if the condensed water is returned to the boiler, the drip is arranged accordingly.

But it is the best practice to lay the main pipe with the lowest part at the boiler, so that the drip will take care of itself, and not require an extra trap, nor interfere with the return circulation.

When steam is turned into cold pipes the water of condensation gets cold after running a short distance, and if it has to go through a small drip pipe full of frost it will probably be frozen. Then, unless it is followed up with a pail of hot water, the whole arrangement will be frozen and a great many bursted pipes will result. Whenever turning steam on in a system of very cold pipes, only one room should be taken at a time, and a pail of hot water should be handy so that if the pipe becomes obstructed it can be thawed immediately without damage.

When pipes become extensively frozen there is nothing to do but take them out and put in new ones.

POINTS RELATING TO STEAM HEATING.

The manner in which a temperature too low to start rapid combustion in wood in steam pipes, operates in originating a fire is by first reducing the oxide of iron (rust) to a metallic condition. This is possible only under certain external conditions, among them a dry atmosphere. *Just as soon as the air is recharged with moisture, the reduced iron is liable to regain, at a bound, its lost oxygen, and in doing so become red hot.* This is the heat that sets the already tindered wood or paper ablaze.

Fig. 142.

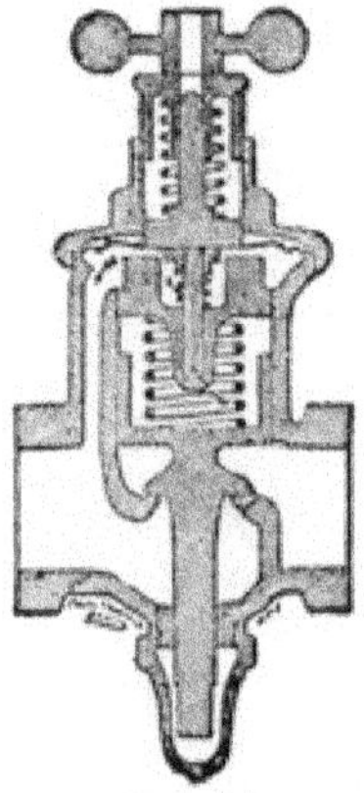

Fig. 143.

Where there is no rust there is no danger from fire with a less than scorching temperature in the pipe or flue. Hence the necessity of keeping steam or hot water fittings in good order.

The indirect system of heating is the most expensive to put in; as to the cost of providing nearly double the heating surface in the coils must be added the cost of suitable air boxes, pipes and registers. For a large installation, this is a serious matter, although for office warming the advantages gained on the score of healthfulness and greater efficiency of employees much more than counterbalance the extra expense.

One horse power of boiler will approximately heat C,000 to 10,000 cubic feet in shops, mills and factories—dwellings require only one horse power for from 10,000 to 20,000 cubic feet.

From seven to ten square feet of radiating surface can be heated from *one square foot of boiler surface,* i. e. the heating surface of the boiler and each horse power of boiler will heat 240 to 360 feet of 1 inch pipe.

POINTS RELATING TO STEAM HEAT.

The profession most nearly related to that of steam engineers is the working steam fitters' occupation. Strictly speaking, the engineer should produce the steam, and it is the steam fitters' place to fix up all the steam pipes and make all the necessary connections: but where the steam plants are small, the engineer may be steam fitter also: hence the introduction in this work of these "Points" which are necessary to be known for the proper care and management of any system of steam or hot water heating.

The care and patience, the mental strain and not infrequently the physical torture incident to fitting up a complicated pipe system cannot adequately be set forth in words.

It is stated to be a fact, that in high pressure hot water heating the water frequently becomes red hot, pressures of 1000 to 1200 pounds per square inch being reached, and when the circulation of the system is defective the pipe becomes visibly red in the dark.

Pipes under work benches should be avoided, unless there is an opening at the back to permit the escape of the heated air, which would otherwise come out at the front.

When both exhaust and live steam are used for heating, many engineers prefer to use independent lines of pipe for each, rather than run the risk of interference and waste caused by admitting exhaust and live steam into the same system at the same time. Nevertheless, the advantages gained by being able to increase the heating power of a system in extremely cold weather by utilizing the entire radiating surface for high pressure steam, are so great that it is probably better so to arrange the system of pipes and connections that this can be done.

Double extra heavy pipe (XX) is used for ice and refrigerating machines (see page 246), as a general rule, makers of this class of machinery obtain but little satisfaction in the use of the ordinary thread joining and use special dies *with uniform taper*—both for couplings, flanges and threading the pipe itself. They do this to protect their reputation and guarantees.

POINTS RELATING TO STEAM HEATING.

Welding boiler and other tubes.—The following is a good way in cases of emergency and can be done on a common forge:

Enlarge one end of the shortest piece, and one end of the long piece make smaller, then telescope the two about ¾ of an inch. Next get an iron shaft as large as will go into the tube and lay across the forge with the tube slipped over it. *Block the shaft up so that the tube will hang down from the top of the shaft.* By such an arrangement the inside of the tube will be smooth for a scraper. When the tube gets to a welding heat strike on the *end* of the short piece first, with a heavy hammer, then with a light and broad-faced hammer make the weld. Borax can be used to good advantage, but it is not necessary. The next thing is to test the tube, which can be done in the following manner: Drive a plug in one end of the tube, stand it up on that end, and fill it with water, if it does not leak the job is well done, if a leak exists the welding must be again done.

SOLID-DRAWN IRON TUBES: CALCULATED BURSTING AND COLLAPSING PRESSURES.

External Diameter.	Thickness.	Internal Diameter.	Bursting Pressure.		Collapsing Pressure.	
			Per Square Inch of Internal Surface.	Per Square Inch of Section of Metal.	Per Square Inch of External Surface.	Per Square Inch of Section of Metal.
Inches.	Inch.	Inches.	Lbs.	Tons.	Lbs.	Tons.
1¼	.083	1.084	7700	22.4	6500	21.7
1⅜	.083	1.209	6900	22.4	5800	21.3
1½	.083	1.334	6200	22.4	5200	21.0
1¾	.083	1.584	5300	22.4	4300	20.3
2	.083	1.834	4500	22.4	3700	19.7
2¼	.095	2.060	4600	22.4	3600	19.0
2½	.109	2.282	4800	22.4	3600	18.3
2¾	.109	2.532	4400	22.4	3100	17.7
3	.120	2.760	4300	22.4	3000	17.0
3½	.134	3.232	4200	22.4	2700	15.7
3¾	.134	3.482	3900	22.4	2400	15.0
4	.134	3.732	3600	22.4	2100	14.3
4½	.134	4.232	3200	22.4	1700	13.0
4¾	.134	4.482	3000	22.4	1600	12.3
5	.134	4.732	2800	22.4	1400	11.7
5½	.148	5.204	2800	22.4	1200	10.3
6		5.704	2600	22.4	1000	9.0

VENTILATION.

The quantity of air for each minute for one person is from four to fifteen feet—and from one-half to one foot should be allowed for each gas jet or lamp.

Heated air cannot be made to enter a room unless means are provided for permitting an equal quantity to escape, and the best places for such exit openings is near the floor.

For healthful ventilation the indirect system of steam heating is by far the best yet devised, for it not only warms the room, but insures perfect ventilation as well. In this system, the air for warming the room is introduced through registers, having first been heated by passing over coils of pipe or radiators suitably located in the air ducts. There is a large volume of pure air constantly entering the room, which must displace and drive out an equal quantity of impure air. This escapes principally around the doors and windows, so that not only is the ventilation effected automatically without the use of special devices, but all disagreeable indraft of cold air is prevented.

One of the cheapest and best methods of ventilation is to have an opening near the floor, opening directly into the flue, or some other outlet especially constructed for it, *with hot water or steam pipes in this opening.* A moderate degree of heat in these pipes will create a draft, and draw out the bad air. Only a few of these pipes are necessary, and the amount of hot water or steam required to heat them is too small to be worthy of consideration.

The use of a small gas-jet, burning continuously, in a pipe or shaft has been found to be a most admirable method of ventilating inside rooms, closets and similar places where foul air might collect if not replaced by fresh. The following table exhibits the result of careful experiments made by Mr. Thomas Fletcher, of England, with a vertical flue 6 inches in diameter and 12 feet high:

TABLE.

Gas Burnt per Hour.	Speed of Current per Minute.	Total Air Exhausted per Hour.	Air Exhausted per Cubic foot of Gas Burnt.	Temperature at outlet. Normal 62° Fahr.
Cubic Feet.	Feet.	Cubic Feet.	Cubic Foot.	
1	205	2,460	2,460	82°
2	245	2,940	1,470	92°
4	325	3,900	975	110°
8	415	4,980	622	137°

EXHAUST STEAM HEATING.

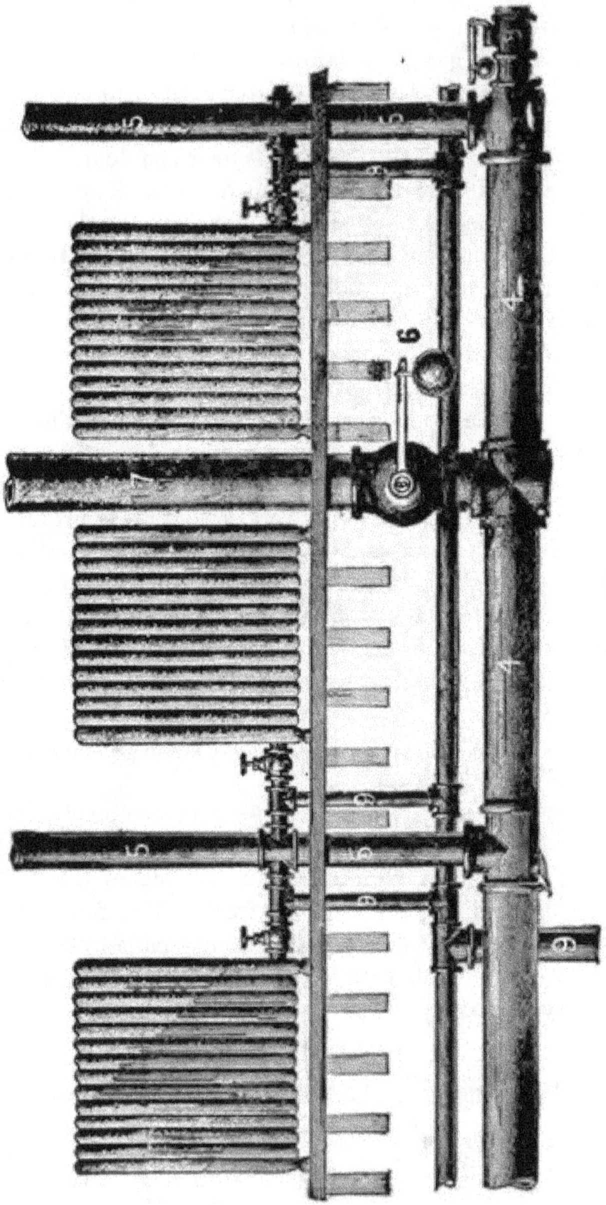

Fig. 144.

VENTILATION.

Taking the experiments as a whole, it will be seen that in a flue 6 inches in diameter, the maximum speed of current which can be obtained with economy is about 200 feet per minute; and this was realized with a gas consumption of 1 cubic foot per hour—1 cubic foot of gas removing 2,460 cubic feet of air.

It should, however, not be required of any system of heating to more than aid in ventilation. It is the architect's or builder's performance to so arrange lower and upper openings to drive out the bad air.

HEATING BY EXHAUST STEAM.

There are two methods of warming by steam heat—one with live steam direct from the boiler, and the other with exhaust steam. These two are frequently carried out in combination, and in fact generally so where exhaust steam is used at all for warming.

In nearly all manufacturing establishments, office buildings, etc., the exhaust steam produced will very nearly, if not quite supply sufficient exhaust steam to furnish all the heat required for heating the building during average weather, although in extremely cold weather, a certain amount of live steam might be necessary to use in connection with the exhaust to supply the required amount of heat.

A simple and convenient device operating upon the suction principle has been found to be most efficient. By this the exhaust steam is drawn almost instantly through the most extensive piping; preventing condensation, freezing, and hammering, after which it is condensed and purified, and fed back into the boiler by the means of a reciprocating pump.

It is claimed that a given quantity of exhaust steam can be circulated by this vacuum system and uniformly distributed through double the amount of heating pipes than could be accomplished by the same quantity of exhaust steam when forced into the heating system by pressure.

Fig. 144 is a well-tried system of heating by exhaust steam in which "7" represents the steam exhaust pipe, with "6" showing back pressure valve with weight to adjust amount of back pressure; "4" "4" are steam supply pipes to radiators; "5" "5" are risers; "9" "9" are condensation return pipes

HEATING BY EXHAUST STEAM.

from the ra liators; "8" is the pressure regulating valve from the boilers. Fig. 144 may also be said to represent the general method of piping used in steam and hot water heating, which is difficult of illustration owing to the fact that each locality where it is used requires a different adaptation.

CARE OF STEAM FITTINGS.

Many steam fittings are lost through carelessness, particularly in taking down old work, but the great bulk are simply "lost" for lack of method in caring for them. This task properly falls upon the engineer, as he usually is entrusted with the selection and ordering of the necessary work. A great saving in the bill of "findings" can be effected by proper attention.

The same systematic care exercised over the other fittings tools, appliances, oil, fuel, etc., used or consumed in the engine and boiler room may be urged with equal emphasis.

							¼ and ⅜ in.
							½ in.
							1 in.
							1¼ in.
							1½ in.
Elbows	Tees.	Nipples.	Plugs.	Reducers.	R's and L's.	Unions	2 in. couplings.

Fig. 145.

Fig. 145 shows a case for keeping fittings, which will enable one to find any particular piece without a moment's delay. In this admirable arrangement it will be seen that the heavy fittings are all at the bottom, the light ones at the top. In the top row of all, the one-quarter and three-eighth inch fittings are placed, being so small that a partition may be put into that

CARE OF STEAM FITTINGS.

row of boxes, and then have plenty of room, and giving twice the capacity to that row of pigeon holes.

Above this case, which is built of one-inch boards, may be put a set of four cupboards, double doors being fitted to each, and thus making a door over each compartment in the fitting rack. The shelves run through these cupboards from end to end, and are not divided by vertical partitions. The necessary brass fittings are kept on these shelves, and the doors are secured by good locks. The lightest fittings are placed on the lower shelves in this cupboard, being in greatest demand.

TOOLS USED IN STEAM FITTING.

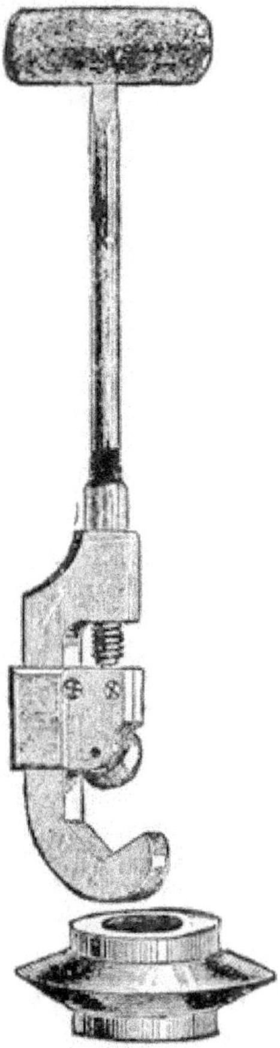

Fig. 146.

Fig. 146 represents one form of a pipe cutter which is made to use by hand; cutters are also made for use by power, which are capable of cutting off pipes of immense size. In an engineers outfit of steam fitting tools 2 sets are advisable—one to cut pipe ⅛th inch to 1 inch, and the other to cut 1 to 2 inch pipe. Figs. 147, 148, represent different forms of pipe tongs—the former called "chain" tongs which will readily hold three inch pipe. Fig. 149 represents a steam fitter's vice which will "take" say, 2½ inch pipe down to ⅛th. Fig. 150 shows a set of taps and dies for small bolts and nuts which is ordinarily to be found in a steam fitter's outfit although used very generally by machinists and others. Fig. 151 shows a pair of gas-pliers which are used by steam fitters in gas pipe jobs. Fig. 152 exhibits the old fashioned alligator wrench.

In ice and refrigerating jobs of pipe fitting special tubes are used to assure a niceness of joints and fitting which is not called for in steam and water service.

TOOLS USED IN STEAM FITTING.

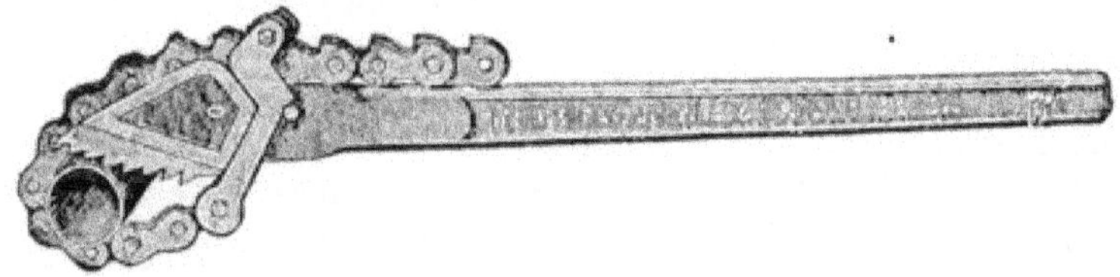

Fig. 147.

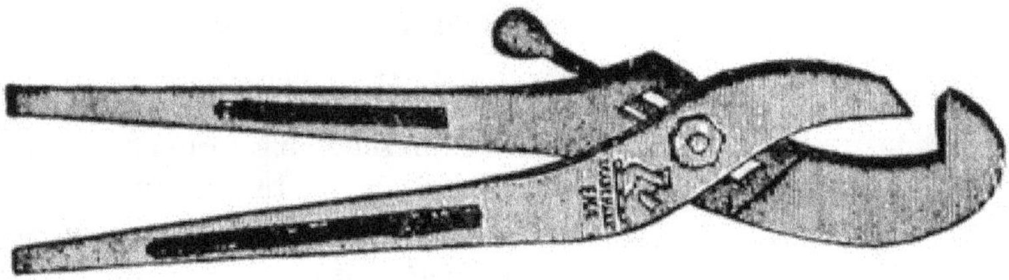

Fig. 148.

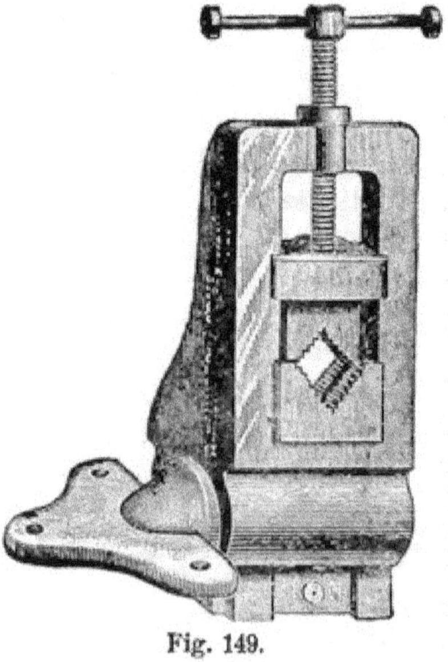

Fig. 149.

COCKS.

The first means in the earliest times of steam engineering, for opening and shutting the passages in the pipes of steam engines were cocks; and these were all worked by hand and required close attention. A boy named Humphry Potter being in charge of one of the cocks of Newcomer's pumping-engines, and desiring time for play, it is said, managed to fasten the lever-handles of the spigots by means of rods and string to the walking beam of the engine, so that each recurrent motion of the beam effected the change required. This was the first automatic valve-motion.

TOOLS USED IN STEAM FITTING.

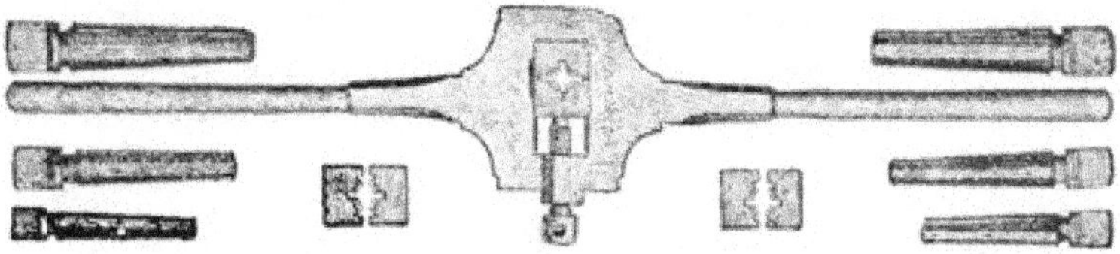

Fig. 150.

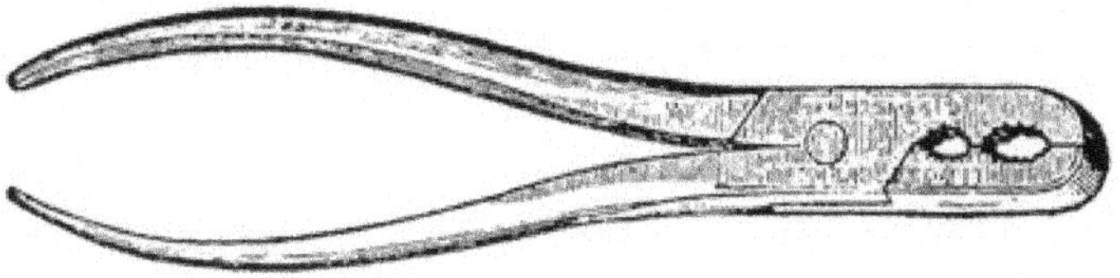

Fig. 151.

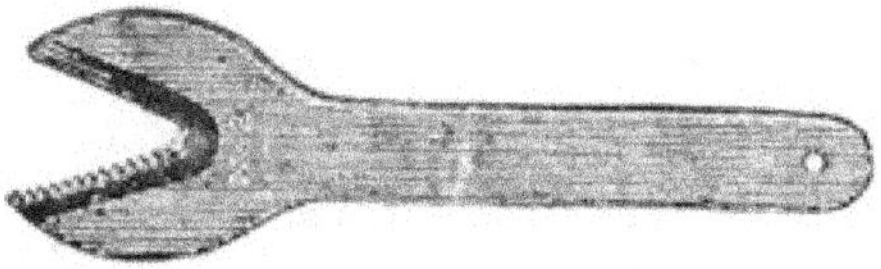

Fig. 152.

VALVES.

The valve is any device or appliance used to control the flow of a liquid, vapor or gas, through a pipe, outlet or inlet in any form of vessel. In this sense the definition includes air, gas, steam, and water cocks of any kind.

The bellows was probably the first instrument of which they formed a part. No other machine equally ancient can be pointed out in which they were required.

By far the most important improvement on the primitive bellows or bag was the admission of air by a separate opening —a contrivance that led to the invention of the valve, one of the most essential elements of steam, of water, as well as pneumatic machinery.

Valves and Cocks.—Generally described, a valve is a lid or cover to an opening, so formed as to open a communication in one direction and close it in another by lifting, turning, or sliding—among the varieties may be classed as, the cock, the slide-valve, the puppet-valve and the clack-valve. A common form of this valve is shown in Fig. 139, page 261.

VALVES AND COCKS.

An every day example of a valve, and almost the simplest known, is that of an ordinary pump where the valve opens upward to admit the water and closes downward to prevent its return.

A valve has a seat, whether it be a gate or circular valve, and is generally turned by a circular handle fitted to the spindle.

Difference between a cock and valve.—The cock is a valve, but a valve is not a cock; the cock is a conical plug slotted and fitted with a handle for turning the cone-shaped valve, with its opening in line, or otherwise, with the opening of the pipe.

Globe Valve is a valve enclosed in a globular chamber, Fig. 135. This, like many other valves, takes its name from its shape.

Globe valves, whenever possible, should be placed *so that the pressure comes under the valve,* or at the side, for if the valve should become loose from the stem (which they often do) if the pressure is on top, there would be a total stoppage of the steam.

Relief Valve is a valve so arranged that it opens outward when a dangerous pressure or shock occurs; a valve belonging to the feeding apparatus of a marine engine, through which the water escapes into the hot well when it is shut off from the boiler.

Hinged Valves constitute a large class, as for example the butterfly valve, clack-valves, and other forms in which the leaf or plate of the valve is fastened on one side of the valve seat or opening.

Valve-bracket is a bracket fitted with a valve.

The Valve-chamber is where a pump valve or steam valve operates.

Valve-cock.—A form of cock or faucet which is closed by dropping of a valve on its seat.

Valve-coupling is a pipe coupling containing a valve.

Valve-seat is the surface upon which a valve rests.

COCKS AND VALVES.

Back pressure valves are ball or clack valves in a pipe which instantly assume the seat when a back pressure occurs. They are illustrated in "6," Fig. 144. Their name signifies their use—to maintain a constant back pressure in heating systems.

Ball valve—a faucet which is opened or closed by means of a ball floating in the water. It constitutes an automatic arrangement for keeping the water at a certain level.

Bib-cock—a faucet having a bent-down nozzle.

Check-valve—a valve placed between the feed pipe and the boiler to prevent the return of the water, etc.

Brine-valve—a valve which is opened to allow water saturated with salt to escape. In marine service it is "a blow-off valve."

Ball valve—a valve occupying a hollow seat. These valves are raised by the passage of a fluid and descending are closed by gravity.

Angle valve is one which forms part of an angle, see Fig. 137.

The double-seat valve or double-beat valve presents two outlets for the water. In the Cornish steam engine this is called the *equilibrium-valve,* because the pressure on the two is very nearly equalized.

Three-way cock is one having three positions directing the fluid in either of three directions. This is illustrated in Fig. 138. The *three-way valve* is also illustrated on page 259, Fig. 136.

Four-way cock is one having two separate passages in the plug and communicating with four pipes.

Gate valve—a valve closed by a gate. This is illustrated in Fig. 140.

Swing or straight-way valve—this is shown in Fig. 141, page 261.

Throttle Valve.—This is the valve used to admit steam to the engine and so termed to distinguish it from the main stop valve located near the boiler—to throttle means to choke—hence the throttling of the steam.

Rotary Valves are those in which the disc, or plug, or other device used to close the passage, is made to revolve for opening or closing, the common stop cock being an illustration.

COCKS AND VALVES.

Lifting Valves are those in which the full cone or stopper is lifted from the valve seat by pressure from below, the poppet, and safety valves being examples.

Pressure regulator valve—this is sometimes called a reducing valve and is illustrated in Figs. 142, 143, on page 262. It is designed to reduce the pressure from a high point in the boiler to a lower one in a system of piping, etc.

Usually the smaller valves, not exceeding 1¼ inch in diameter, are wholly of gun-metal; the larger are commonly made with cast-iron bodies and gun-metal fittings. The smallest valves, from ¼ up to ½ inch inclusive, have the disk solid with the spindle, and have an ordinary stuffing-box with external gland. Valves of ¾ inch and upwards have the disk loose from the spindle; up to 3 inch valves the spindles are screwed to work inside the casing; above that size the screwed portion is outside the casing. Above the 3-inch size the nozzles of the cast-iron bodies are generally flanged instead of tapped.

STEAM FITTINGS.

A few of the principal sorts have been illustrated in this work and still others will be described in the "Index" at the close of the work.

Fig. 123, page 251, illustrates *an elbow* with outlet. This is sometimes spelled with the capital L, and again as an ell.

Fig. 124 shows a long *nipple*.

Fig. 125, page 253, exhibits a *bushing*, used to reduce one size pipe in a line to another.

Fig. 126 is a *cross tee*. This is frequently spelled with a capital T.

Fig. 127 is *a plug*—used to stop apertures in plates or pipes.

Fig. 128, page 254, illustrates *a lock nut*.

Fig. 129 shows a T, as illustrating the difference between a T and a cross T, Fig. 126.

Fig. 130 is *a coupling*.

Fig. 131, page 255, represents *a reducing coupling*.

Fig. 132 is an illustration of a pipe *union*.

Fig. 133 is a plain *elbow* (see also Fig. 123.)

STEAM PIPE AND BOILER COVERINGS.

This subject relates to the *radiation of heat*, which allows a reference to the laws of heat and tables of radiating power of various substances, as set forth on pages 212, 215.

The importance of a protection of exposed surfaces from radiation of heat is now undisputed, and many experiments have determined very closely the relative value of the various non-conducting substances.

Table of the CONDUCTING POWER *of various substances.*

Substance.	Conducting Power.	Substance.	Conducting Power.
Blotting Paper.........	.274	Wood, across fibre....	.83
Eiderdown....	.314	Cork	1.15
Cotton or Wool, any density............	.323	Coke, pulverized......	1.29
		India Rubber.	1.37
Hemp, Canvas	.418	Wood, with fibre....	1.40
Mahogany Dust.........	.523	Plaster of Paris.......	3.86
Wood Ashes............	.531	Baked Clay...........	4.83
Straw....	.563	Glass............	6.6
Charcoal Powder.......	.636	Stone.	13.68

By the above table may be judged the comparative value of different coverings; blotting paper with *its confined air*, standing at one end of the list, stone at the other. It should be noted that *the less the conducting power the better protection against radiation.*

A non-conducting coating for steam pipes, etc., used for many years with perfect satisfaction, can be prepared by any steam user. It consists of a mixture of wood sawdust with common starch, used in a state of thick paste. If the surfaces to be covered are well cleaned from all trace of grease, the adherence of the paste is perfect for either cast or wrought iron; and a thickness of 1 inch will produce the same effect as that of the most costly non-conductors. For copper pipes there should be used a priming coat or two of potter's clay, mixed thin with water and laid on with a brush. The sawdust is sifted to remove too large pieces, and mixed with very thin starch. A mixture of two-thirds of wheat starch with one-third of rye starch is the best for this purpose. It is the common practice to wind string spirally round the pipes to be treated to

PIPE AND BOILER COVERINGS.

secure adhesion for the first coat, which is about 1-5th of an inch thick. When this sets, a second and a third coat are successfully applied, and so on until the required thickness is attained. When it is all dry, two or three coats of coal tar, applied with a brush, protect it from the weather.

A very efficient covering may be made as follows: 1, wrap the pipe in asbestos paper—though this may be dispensed with; 2, lay slips of wood lengthways, from 6 to 12 according to size of pipe—binding them in position with wire or cord; 3, around the framework thus constructed wrap roofing paper, fastening it by paste or twine. For flanged pipe, space may be left for access to the bolts, which space should be filled with felt. Use tarred paper—or paint the exterior.

While a very efficient non-conductor, hair or wool felt has the disadvantage of becoming soon charred from the heat of steam at high pressure, and sometimes taking fire. The following table, prepared by Chas. E. Emory, Ph. D., shows *the value* of various substances, taking wool felt as a *unit*.

TABLE OF RELATIVE VALUE OF NON-CONDUCTORS.

Non-Conductor.	Value.	Non-Conductor.	Value.
Wood Felt	1.000	Loam, dry and open	.550
Mineral Wool No. 2	.832	Slacked Lime	.480
Do. with tar	.715	Gas House Carbon	.470
Sawdust	.680	Asbestos	.363
Mineral Wool No. 1	.676	Coal Ashes	.345
Charcoal	.632	Coke in lumps	.277
Pine Wood, across fibre	.553	Air space, undivided	.136

LINEAR EXPANSION OF STEAM PIPES.

Wrought iron is said to expand 1-150,000 of an inch for each degree of heat communicated to it; to make the calculation take the length of the pipe in inches, multiply it by the number of degrees between the normal temperature it is required to attain when heated, and divide this by 150,000. Suppose the pipe is 100 feet long, and its temperature zero, and it is desired to use it to carry steam at 100 pounds pressure—equal to a temperature of 338 degrees - multiply 100 feet by 12 to reduce it to inches, and by 338, the difference in temperature; divide

LINEAR EXPANSION OF STEAM PIPES.

this by 150,000, and the result will be 2.7 inches, which would be the amount of play that would be required, in this instance, in the expansion joint.

Figs. 153 and 154 show a properly designed arrangement of steam connections for a battery of boilers. To the nozzles, risers are attached by means of flanges, and from the upper ends of these risers pipes are led horizontally backwards into the main steam pipe. In this horizontal pipe, the stop valves, one to each boiler, are placed. These valves should have flanged ends as shown, so that they may be easily removed, if repairs become necessary, without disturbing any other portion of the piping. Unlike the engraving, the valve C should be arranged in another position: the stem should, of course, be horizontal or nearly so, in order that the valve may not trap water.

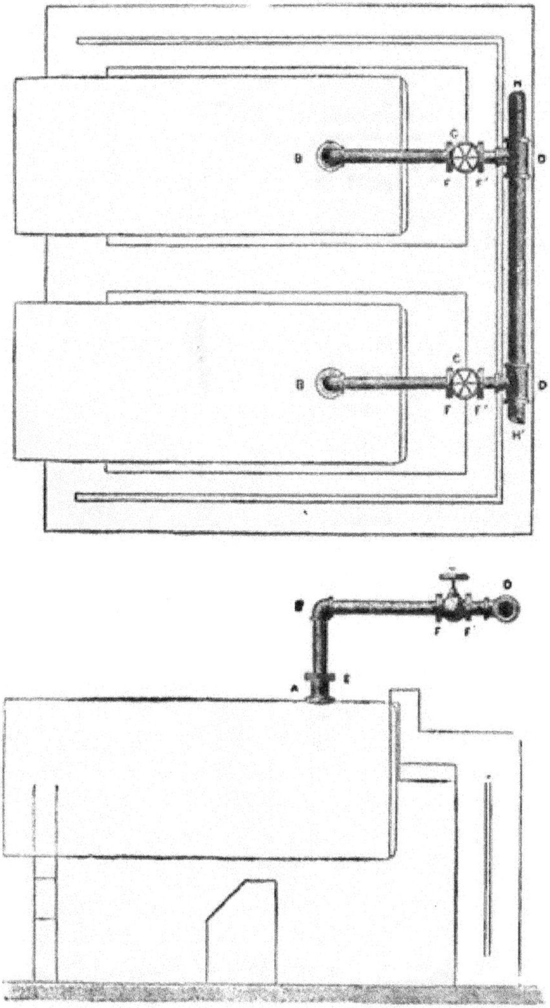

Figs. 153 and 154.

LINEAR EXPANSION OF STEAM PIPES.

By this arrangement it will be seen that the movements of the boilers and the piping itself are compensated for by the spring of the pipes. The height of the risers should never be less than three feet, and when there are eight or ten boilers in one battery, they should be, if room permits, six to eight feet high, and the horizontal pipes leading to main steam pipe should be ten or twelve feet or more.

THE STEAM LOOP.

This is an attachment to a steam boiler, designed to return water of condensation. It invariably consists of three parts, viz.: the "riser," the "horizontal," and the "drop leg," and usually of pipes varying in size from three-fourth inch to two inches. Each part has its special and well defined duties to perform, and their proportions and immediate relations decide and make up the capacity and strength of the system. It is, in fact, nothing but a simple return pipe leading from the source of condensation to the boiler, and, beyond this mere statement, it is hardly possible to explain it; it has, like the injector and the pulsometer pump been called a paradox.

The range of application of the steam loop practically covers every requirement for the return of water of condensation. If used in connection with a steam engine, pump, etc., a separator of any simple form is connected in the steam pipe as close as possible to the throttle. From the bottom of the separator the loop is led back to the boiler, and the circulation maintained by it will dry the steam before it is admitted to the cylinder.

There is necessary to its operation a slight fall in temperature at the head of the loop, which is accompanied by a corresponding fall in pressure. The water accumulating in the lower end of the loop next to the separator, as soon as it fills the diameter of pipe, is suddenly drawn or forced to the horizontal by that difference in pressure. It is immaterial how far the water has to be taken back, or how high it is to be lifted. There is one system now in daily operation lifting the condensed water over thirty-nine feet, and another lifting it over

THE STEAM LOOP.

sixty-three feet. The strength of the system is increased by length and height, the only limit to its operation being the practicability of erecting the necessary drop leg, the height of which depends on difference in pressures.

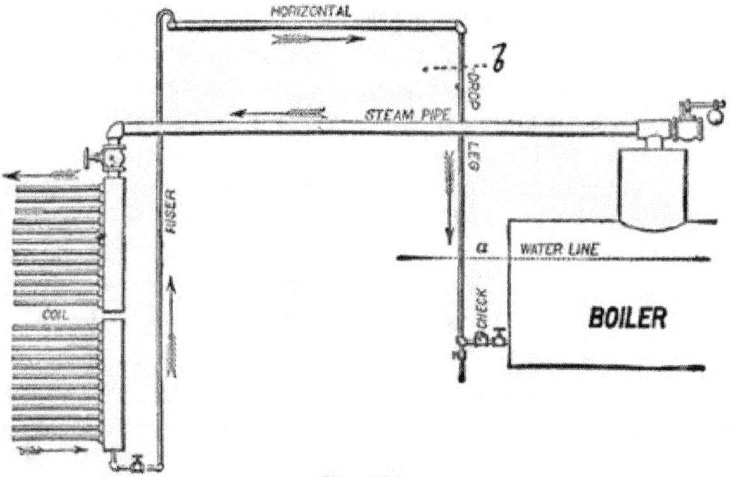

Fig. 155.

Fig. 155 is an illustration of its application to a radiating coil. To understand the philosophy of its action, and referring to the illustration, let us assume that all the valves are open, and full boiler pressure is freely admitted throughout the steam pipe, coil and loop. Now, if the pressure were exactly uniform throughout the whole system, the water in the loop would stand at *a* on the same level as the water in the boiler. But, as a matter of fact, the pressure is not uniform throughout the system, but steadily reduces from the moment of leaving the dome. This reduction of pressure is due in part to condensation, and in part to friction, and although generally small, is always present in some degree. The pressure may be intentionally reduced at the valve on the coil, and reduction necessarily results from condensation within the coil itself. A still further reduction takes place through the loop, so that the lowest pressure in the whole system will be found at *a*, the point in the loop furthest from the boiler, reckoned by the flow of steam

THE STEAM LOOP.

Now it is known that water of condensation invariably works towards, and accumulates in, a "dead end." This is due to the fact that, as already shown, the pressure is lower at the "dead end" than at any other point in the system, and, as a consequence, there is a constant flow, or sweep, of steam towards the point of least pressure, which flow continues as long as condensation goes on. This sweep of steam carries along with it all the water formed by condensation or contained in the steam, at first in the form of a thin film swept along the inner surface of the loop, and afterwards, when the accumulation of water is sufficient, in the form of small slugs or pistons of water, which completely fill the pipe at intervals, traveling rapidly towards the dead end. The action of the steam sweep is vastly more powerful than is usually supposed, and, of course, operates continuously and infallibly to deposit the water in the dead end as fast as accumulated.

In practice, water will speedily be carried over by the loop and accumulate in the drop leg until it rises to the level *b*, which would balance the difference in pressure. As the loop will still continue to bring over water, it follows that as fast as a slug or piston of water is deposited by the steam on the top of the column at *b*, *it overbalances the equilibrium and an equal amount of water is discharged from the bottom of the column through the check valve into the boiler.*

The result of the practical operation of many systems of this ingenious device show advantages as follows:

1. Return of pure water to the boiler and saving the heat contained in said water.

2. Preserving more uniform temperatures, thus avoiding the dangers due to expansion and contraction.

3. Prevention of loss from open drains drips, tanks, etc.

4. Maintaining higher pressure in long lines of piping, in jackets, driers, etc.

5. Enabling engines to start promptly.

6. Saving steam systems from water, thereby reducing liability to accident.

BOILER MAKERS' TOOLS AND MACHINERY.

Fig. 156 represents a pair of jack screws. These are invaluable devices for use in boiler-shops, and also in establishments where ponderous machinery has to be shifted or otherwise handled.

Fig. 156.

But few machine tools are used in making steam boilers, and they are generally as follows:

1st.—*The Rolls*, operated either by hand levers or power; used for bending the iron or steel plates into circular form.

2d.—A wide *power planer* for trimming the edges of the sheet perfectly straight and true.

3d.—*Heavy Shears* for trimming and cutting the plates.

4th.—A *Power Punch* for making the rivet holes.

5th.—A *Disc* for making the large holes in the tube sheets to receive the ends of the tubes.

6th.—*Rivet heating furnaces* and frequently *steam riveting machines*.

The hand tools needed by boiler makers are equally few, consisting of *riveting hammers* and hammers for striking the chisels, *tongs* to handle hot rivets, *chipping chisels* used in trimming the edges of plates, *cape chisels* for cutting off iron or making holes in the sheets, *expanders* to set the tubes, and also *drift pins* to bring the punched sheet exactly in line.

Fig. 157 exhibits an improved pattern of the well-known tool—dudgeon expander.

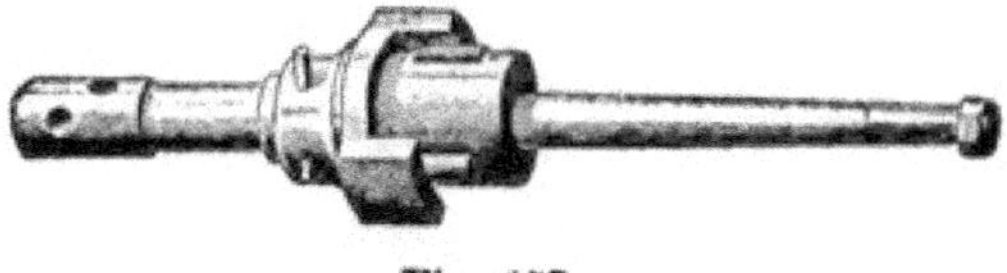

Fig. 157.

STEAM.

Steam is water in a gaseous state; the gas or vapor of water; it liquifies under a pressure of 14.7 and temperature of 212° F.

Steam is a joint production of the intermingling of water and heat. Water is composed of two gases which have neither color nor taste, and steam is made up of the same two gases with the addition only of that mysterious property called heat by which the water becomes greatly expanded and is rendered invisible. The French have a term for steam which seems appropriate when they call it water-dust.

This is what takes place in the formation of steam in a vessel containing water in free communication with the atmosphere. At first, a vapor is seen to rise that seems to come from the surface of the liquid, getting more and more dense as the water becomes hotter. Then a tremor of the surface is produced, accompanied by a peculiar noise which has been called *the singing* of the liquid; and, finally, bubbles, similar to air bubbles, form in that part of the vessel which is nearest to the fire, then rise to the surface where they burst, giving forth fresh vapor.

The curious fact must be here noted that if water be introduced into a space entirely void of air, like a vacuum, it vaporizes instantaneously, no matter how hot or cold, so that of an apparent and fluid body there only remains an invisible gas like air.

That steam is *dry* at high pressure is proved by an experiment which is very interesting. If a common match head is held in the invisible portion of the steam jet close to the nozzle, it at once lights, and the fact seems convincing as to complete dryness, as the faintest moisture would prevent ignition even at the highest temperature. This experiment proves dryness of the steam at the point of contact, but if throttling exists behind the jet, the steam supplied by the boiler may be in itself wet and dried by wire drawing.

Dead steam is the same as exhaust steam.

Live steam is steam which has done no work.

Dry steam is saturated steam without any admixture of mechanically suspended water.

STEAM.

High-pressure steam is commonly understood to be steam used in high-pressure engines.

Low-pressure steam is that used at low pressure in condensing engines, heating apparatus, etc., at 15 lbs. to the inch or under.

Saturated steam is that in contact with water at the same temperature; saturated steam is always at its condensing point, which is always the boiling point of the water, with which it is in contact; in this it differs from superheated steam.

Superheated steam, also called steam-gas, is steam dried with heat applied after it has left the boiler.

Total heat of steam is the same as steam heat.

Wet steam, steam holding water mechanically suspended, the water being in the form of spray.

Specific gravity of steam is .625 as compared to air under the same pressure.

The properties which make it so valuable to us are:

1. The ease with which we can condense it.

2. Its great expansive power.

3. The small space in which it shrinks when it is condensed either in a vacuum chamber or the air.

A cubic inch of water turned into steam at the pressure of the atmosphere will expand into 1,669 cubic inches.

WATER HAMMER.

The fact that steam piping methods have not kept pace with the demands of higher pressures and modern practice is evidenced by the increasing number of accidents from the failure of pipes and fittings.

There has not been, for the rapid increase of pressure used, a proportionate increase in strength of flanges, number and size of bolts used, and more generous provision for expansion and contraction. Valves and fittings also require greater attention in their design, construction and manipulation.

WATER HAMMER.

It is well known that the presence of condensed water in pipes is a source of danger, but little is known of what exactly goes on in the pipe. We have the incompressible liquid, the expansive gas, and the tube with a "dead head" or dead end as it is called, or where the end of the pipe is closed. Seeing that the tube or pipe is capable of withstanding all the pressure that the steam can give, it is difficult to account for the tremendous repelling force, which is, undoubtedly, brought into operation in explosions or ruptures of steam pipes carrying what are now comparatively low pressures.

The cause of the bursting is undoubtedly *water hammer or water ram*, which accompanies large, long steam pipes, filled with condensed water.

If steam be blown into a large inclined pipe full of water, it will rise by difference of gravity to the top of the pipe, forming a bubble; when condensation takes place, the water below the bubble will rush up to fill the vacuum, *giving a blow directly against the side of the pipe.* As the water still further recedes the bubble will get larger, and move farther and farther up the pipe, the blow each time increasing in intensity, for the reason that the steam has passed a larger mass of water, which is forced forward by the incoming steam to fill the vacuum. The maximum effect generally takes place at a "dead end."

In fact, under certain conditions, a more forcible blow is struck when the end of the pipe is open, as, for instance, when a pipe crowned upward is filled with water, one end being open and the steam introduced at the other. A bubble will in due time be formed at the top of the crown, when the water will be forced in by atmospheric pressure from one end and by steam pressure from the other, and the meeting of the two columns frequently ruptures the pipe.

The remedy for this is simple, the pipes must be properly located so as to drain themselves or be drained by rightly located drip cocks. The drip should be the other side of the throttle valve, and if steam is left on over night this valve should be left open enough to drain out all the water.

HAZARDS OF THE BOILER ROOM.

Where there is great power, there is great danger.

When the pressure is increased, the danger is increased.

When the pressure is increased, diligence, care and scrutiny should be increased.

During the twelve years between 1879 and 1891 there were recorded 2,159 boiler explosions; these resulted in the death of 3,123 persons, and in more or less serious injury to 4,352 others. Besides these there were innumerable other accidents during the same period, caused by other means, which emphasizes the gravity of this cautionary "chapter of accidents."

Every boiler constructed of riveted plate and carrying a high head of steam, holds in constant abeyance, through the strength of a disruptive shell, a force, more destructive in its escaping violence than burning gunpowder. To the casual observer there is no evidence of this; and it is only when a rupture takes place of such a character as to liberate *on the instant the entire contents of the boiler* that we get a real demonstration of the fact. Unfortunately a steam boiler never grows stronger, but deteriorates with every day's age and labor, subjected, as it is, to all sorts of weakening influences; and fractures often occur, which, if not at once repaired, would speedily reduce the strength of the boiler to the point of explosion.

In the case of a boiler we have, first, a vessel of certain strength, to resist strains; and second, expansive steam and water contained therein. It must be plain that if the strength of the vessel is superior to the internal pressure there can be no explosion, and also, on the contrary, if we allow the pressure to go above the strength of the vessel, that there must be a rupturing and an explosion, but it will be in the weakest place of that vessel.

Experiments by the most eminent men have failed to discover any mysterious gas formed by boiling water, or by any mixture

STEAM BOILER EXPLOSIONS.

of air and water. Boilers have been built for the express purpose of trying to explode them under various conditions of high and low water, and nothing in regard to the sudden generation of any gas has been discovered. Again, disastrous explosions that have occurred have been of vessels that contained no water and were not in contact with fire, flame or heated air, but were supplied by steam some distance away.

The destructive efforts of the vaporization attendant upon explosions seem to be due to the subsequent expansion of the steam so formed, rather than to the intensity of its pressure; low or high steam *alone* has very little to do with boiler explosions; nor high or low water necessarily.

The one great cause of boiler explosion is the inability of the boiler to withstand the pressure to which it is subjected at the time, and this may be brought about by any one of the following causes, viz.:

1. Bad design, in which the boiler may not be properly strengthened by stays and braces; deficient water space, preventing the proper circulation of the water.

2. Bad workmanship, caused by the punching and riveting being done by unskilled workmen.

3. Bad material, blisters, lamination, and the adhesion of sand or cinders in the rolling of the plate.

4. By excessive pressure, caused by the recklessness of the engineer, or by defective steam-gauges or inoperative safety-valves.

5. Overheating of the plates, caused by shortness of water. When water is poured on red-hot surfaces it does not touch the surface, but remains in the spheroidal state at a little distance from it, being apparently surrounded by an atmosphere of steam. It assumes this state above 340°; when the temperature falls to about 288° it touches the surface and commences boiling.

STEAM BOILER EXPLOSIONS.

6. By accumulation of scale, mud, or other deposit, which prevents the water gaining access to the iron. This causes the seams to leak, the crown-sheet to bulge or come down.

One is unable to find any proof that boilers do generally explode at about starting time, nor is that statement, to the best of information, founded on any basis of fact, but was first affirmed by parties who had designed a boiler especially arranged to avoid that imaginary danger.

No one supposes that inspection will absolutely prevent all explosions; but rigid inspection will discover defects that might end in explosion.

Low water is dangerous from the fact that it leaves parts of the boiler to be overheated and the strength of iron rapidly decreases in such a case. In fact, an explosion caused by low water might be expected to be less disastrous than if the water was higher, other conditions being equal, from the fact of there being less water at a high temperature ready to flash into steam at the moment of liberation.

Testing new boilers *under steam pressure* is both dangerous and unwise—the hot water expansion test is just as efficient, less costly and safe in every respect—hence, there is no occasion for a steam test. A manufacturer was testing a boiler in the way mentioned when a rivet in a brace blew out, and the contents of the boiler rushed out, striking a man in the face, and parboiling him from head to foot. Another who was inspecting the boiler, was struck on the head and enveloped in steam and water; another was also scalded from the shoulders down; another was injured about the arms; a fifth man was scalded and severely injured about the back. The apartment was so filled with steam that the victims could not be rescued until all the damage mentioned had been done to them.

Danger from exploding steam pipes is greater than supposed. An inspector in a pipe works was testing a tube by means of a double-action hydraulic pump; the pipe suddenly burst with

HAZARDS OF THE BOILER ROOM.

the pressure of 5000 pounds to the square inch, and the water striking the unfortunate man on his face, he was killed on the spot.

There is a tendency on the part of engineers to trust too implicitly in their steam gauges. These are usually the only resort for determining the steam pressure under which the boiler may be working. But the best gauges are liable to err, and after long use to require a re-adjustment. It is fortunate, however, that the error is usually upon the safe side of indicating more than the actual pressure.

Any boiler that has been standing idle for a few weeks or months is a dangerous thing to enter, and no one should attempt it until it has been thoroughly ventilated by taking off all the man-hole and hand-hole plates and throwing water into it. This is due to the presence of a gas which is generated from the refuse and mud, or scale, which, to a greater or less degree, remains in all boilers. Contact with fire is certain to result in an explosion. Not long since a locomotive was in a roundhouse, where it had been waiting some weeks for repairs. Some of the tubes were split and a man was pulling them out. He had only removed one or two when, putting in his lamp to see what remained, there was a fearful explosion which shook the shop. There are many other places which are unsafe to enter when they have been long closed, such as wells, pits of any kind, and tanks. Precisely what the nature of the gas is no one seems to know, but it is assuredly settled that a man who goes into it with a light seldom comes out unharmed.

The gas most likely to fill idle boilers in cities is sewer gas, that gets in through the blow-off pipe, which is left open and generally connects with the sewer; hence, the connection with the sewer by the blow-off pipes should receive attention.

Boilers are sometimes unexpectedly emptied of their contents by the operation of the principle of the syphon; a boiler is so piped that a column of water may be so formed as to draw out of the boiler its entire contents. Danger ensues if this is done while the boiler is being fired.

FUEL OIL.

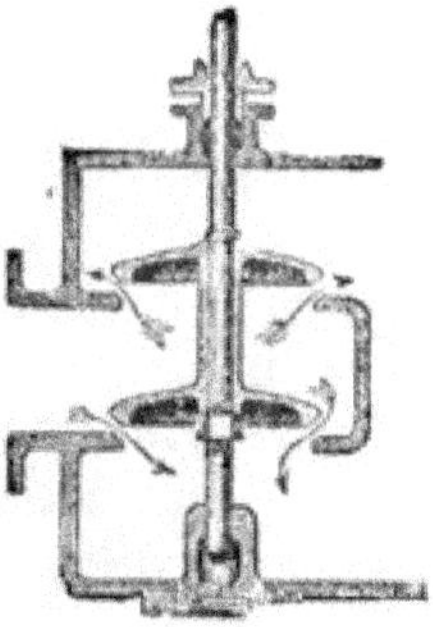

The long experimental use of petroleum or natural oil as a combustible has developed but one serious objection to its wide spread and popular adoption; that objection arises from its liability to ignite and cause destruction by fire; but

THE HAZARDS OF FUEL OIL may be remedied by the observance of the following rules adopted by a certain fire underwriters' association:

"Vault to be located so that the oil it contains can burn without endangering property and have a capacity sufficient tc hold twice the entire quantity of oil the tanks within can contain.

Location of vault to be left to the approval of the Superintendent of Surveys. Distance from any property to be regulated by size of tank.

Vaults to be underground, built of brick, sides and ends to be at least 16 inches thick and to be made water tight with hydraulic cement; bottom to be water tight, concrete, dished toward centre, and inclined to one end so as to drain all overflow or seepage to that end, said incline to be to the end opposite to that from which the tank is to be tapped; top to be supported with heavy iron I-beams, with arches of solid brick sprung from one beam to its neighbors, and to have at least twelve inches of dirt over the masonry.

Vault to be accessible by one or more large man-holes, which, when not in use, are to be kept locked by a large padlock of three or more tumblers, key to be held by some responsible party.

A trough must run from one end of the vault to the other, directly under each tank, and in the same direction as the tank or tanks.

Tank to be of boiler iron or steel, at least 3-16 inch in thickness, to be cold riveted, rivets to be not less than 3-8 inch in

RULES RELATING TO USE OF FUEL OIL.

diameter and not over 1 inch apart between centres; the entire outer surface of tank to have two good coats of coal tar or mineral paint before the tank is placed in position.

No tank shall be over 8 feet in diameter by 25 in length, nor shall any vault have over two tanks.

When tank is set, the bottom of the tank must be 3 inches above the concrete floor of the vault, and must be in saddles of masonry not less than twelve inches in thickness, built from the concrete floor of the vault, said saddles not to be more than 3 feet apart between centres, and laid in hydraulic cement, with an opening through centre for drainage.

Tank must incline 1 inch per 10 feet in length toward the end from which it is to be tapped, said incline of the tank to be opposite to the incline at the bottom of the vault.

The filling pipe, man-hole, telltale or indicator, pump supply connection, steam connection, overflow pipe and ventilating pipes, where they connect with tank, must be made petroleum tight by the use of litharge and glycerine cement.

Flanges to make tank ¾ inch in thickness to be riveted on the inside so as to furnish a satisfactory joint where connections are made, must be used.

Filling pipe connection must have gas-tight valve between the tank and hose coupling, which must be kept closed and locked unless the tank is being filled. Each tank must have ventilating pipes at least 1½ inches in diameter, one of which must connect with one end of the top of the tank and must be in the form of an inverted J, a union to be placed in pipe just below the bend, within which shall be placed a diaphragm of fine wire gauze ; the other ventilating pipe must be at the other end of the top of the tank and must be conducted to the inside of the smoke stack or into the open air at least 10 feet above the surface, so that all the gases that form in the tank will be constantly changed.

Tank must have indicator to show height of oil in tank at all times, said indicator to be so arranged as to allow no escapement of gases from tank.

RULES RELATING TO USE OF FUEL OIL.

All pipes leading from the tank to the pump or place of burning, must incline toward the tank, and have a fall of at least 2 feet from bottom of stand pipe to top of storage tank, and must be so constructed that the feed pipe from stand pipe to burners shall be entirely above burners, so that no pockets of oil can be formed in any one of the pipes between the main tank, stand pipe, oil pump or place of burning.

The vault shall be air tight as near as possible, and must have two ventilating pipes of iron of 4 inches diameter, both inlet and outlet pipes to reach within 6 inches of the bottom of the vault, the outlet ventilating pipe to rise above surface 8 feet, and the inlet ventilating pipe to rise above surface 6 feet.

Syphon to be arranged so as carry out any seepage or leakage into the vault, and discharge same upon the ground, where its burning would not endanger surrounding property."

The following are a part of the rules adopted by the German Government to prevent accidents in mills and factories: they are equally applicable in all places where steam power is used:

"All work on transmissions, especially the cleaning and lubricating of shafts, bearings and pulleys, as well as the binding, lacing, shipping and unshipping of belts, must be performed only by men especially instructed in or charged with such labors. Females and boys are not permitted to do this work.

The lacing, binding or packing of belts, if they lie upon either shafting or pulleys during the operation, must be strictly prohibited. During the lacing and connecting of belts, strict attention is to be paid to their removal from revolving parts, either by hanging them upon a hook fastened to the ceiling, or in any other practical manner ; the same applies to smaller belts which are occasionally unshipped and run idle.

While the shafts are in motion they are to be lubricated, or the lubricating devices examined only when observing the following rules : (1) The person performing this labor must either do it while standing upon the floor, or by the use of (2) firmly

GOVERNMENT RULES TO PREVENT ACCIDENTS.

located stands on steps, especially constructed for the purpose, so as to afford a good and substantial footing for the workman; (3) firmly constructed sliding ladders, running on bars; (4) sufficiently high and strong ladders, especially constructed for this purpose, which by appropriate safeguards (hooks above or iron points below) afford security against slipping.

All shaft bearings are to be provided with automatic lubricating apparatus.

Only after the engineer has given the well-understood signal, plainly audible in the workrooms, is the engine to be started.

If any work other than lubricating and cleaning of the shafting is to be performed while the engine is standing idle, the engineer is to be notified of it, and in what room or place such work is going on, and he must then allow the engine to remain idle until he has been informed by proper parties that the work is finished.

Plainly visible and easy accessible alarm apparatus shall be located at proper places in the workrooms, to be used in case of accident to signal to the engineer to stop the engine at once.

All projecting wedges, keys, set-screws, nuts, grooves or other parts of machinery, having sharp edges, shall be substantially covered.

All belts or ropes which pass from the shafting of one story to that of another shall be guarded by fencing or casing of wood, sheet-iron or wire netting four feet, 6 inches high.

The belts passing from shafting in the story underneath and actuating machinery in the room overhead, thereby passing through the ceiling must be enclosed with proper casing or netting corresponding in heigth from the floor to the construction of the machine. When the construction of the machine does not admit of the introduction of casing, then, at least, the opening in the floor through which the belt or rope passes should be inclosed with a low casing at least four inches high.

GOVERNMENT RULES TO PREVENT ACCIDENTS.

Fixed shafts, as well as ordinary shafts, pulleys and fly-wheels, running at a little height above the floor, and being within the locality where work is performed, shall be securely covered."

The most simple and efficient of all substances for fire extinguishment is sulphur. This, by heat, absorbs oxygen and forms sulphurous acid, the fumes of which are much heavier than the air. The quantity required would be small. Besides sulphur, which gives every satisfaction, both in its effects and from its low cost, we find a similar property in another active and cheap substance, ammonia. An automatic sulphur extinguishing apparatus can be make of various forms.

If night repairs, Sunday, or any other work which requires the use of artificial light (especially portable lights of any kind) becomes necessary, more than one man should be employed, one of whom should be capable of starting the engine or pump instantly in case of fire.

In guarding against explosion it is conceded that the main reliance is to have the boiler made strong enough to stand both the regular load or any unexpected strain caused by the stoppage of the engine; it is also the tendency of the times to proceed towards higher and higher figures in steam pressure, until now it is not unfrequent to see 150 lbs. to the square inch indicated by the gauge; the larger the boiler, also, the more economically it can be run and this, as in the two cases before cited, requires extra precautions in building the boiler with great regard to strength in every part

The following rules posted in a certain factory aremost excellent for their directness.

"Wear close-fitting clothes; have a blouse or jacket to button close around the waist and body; have sleeves to fit arms closely as far up as the elbow; never wear a coat around machinery; never approach a pair of gears or pulley from the driving side; never attempt to save time by potting, or trying to pot on any fast-moving belts without slacking up or stopping entirely to do it. Never allow an inexperienced person to go through the mills without an attendant; never allow a woman

FACTORY RULES FOR PREVENTION OF ACCIDENT.

to go through a mill, no matter how many attendants, while in motion; never attempt to go through the mill in the dark, you may forget the exact location of some dangerous object and seek to avoid it, but it is still there, noiselessly waiting a chance to wreck you; never allow any dangerous place to go unguarded; keep your eye open while oiling; never relax your vigilance for an instant, it may cost you your life. If you feel a gentle tug on your clothes, grab, and grab quick, anything you can cling to, and don't let go till after the clothes do."

WATER CIRCULATION.

Water consists of an innumerable quantity of extremely minute particles called molecules. These particles have the property of being able to glide over, under, and to and from each other almost without resistance or friction. When water is heated in a boiler the action that takes place is this: As the heat is applied, the particles nearest the heated surfaces become expanded or swollen, and are so rendered lighter (bulk for bulk) than the colder particles, they are therefore compelled to rise to the highest point in the boiler.

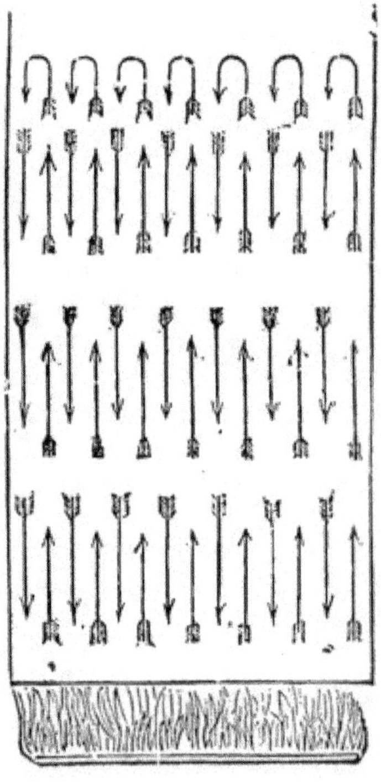

Fig. 158.

This upward action is vividly shown by the illustration on page 242, and by Fig. 158, where the warmer particles are ascending and the cooler ones are descending by a process which is endless so long as heat is applied to the lower part of the containing vessel.

The cause of circulation is the result of an immutable law of nature (the law of gravitation), and is so simple that with

WATER CIRCULATION.

moderate care in its manipulation failures in arranging steam heating apparatus are next to impossible. A very slight experience suffices to show that a pipe taken from the top of a boiler and given a direct or gradual rise to the point furthest from the boiler, and then returned and connected into it at the bottom, will upon the application of heat, cause the water to circulate. It is not necessary that the water should boil or even approach boiling point, to cause circulation, as in a properly constructed apparatus the circulation commences soon after the heat is applied and immediately the temperature is raised in the boiler. It is a very common error to suppose that the circulation commences in the flow or up pipe, whereas it is just the reverse. The circulation is caused by the water in the return pipe and can be described as a stream of heated particles flowing up one pipe from the boiler and a stream of cooler particles flowing down another pipe into the boiler; or it might be described as a means of automatically transporting heated water from the lower to the upper parts of a building, and providing a down flow of cold water to the boiler to be heated in turn.

Those having in charge the erection of hot-water systems for heating buildings, will do well to remember that the circulation they expect depends entirely upon the expansion of particles when heated, and that they must avoid as much as possible friction, exposure of flow pipes to very low temperature, and frequent or numerous short bends.

When properly arranged the action of "the steam loop" is a very good illustration of the circulation of hot water and steam; the flow is continuous, rapid and positive.

NOTE.—When the steam loop is properly connected, the stop valve at the boiler should always be left open and full pressure maintained in the steam pipe over night or over Sunday. The loop will keep up a powerful circulation, returning all water to the boiler as fast as condensed. On starting up in the morning, it is only necessary to open the waste cocks and blow out what little water may have condensed in the cylinders themselves. The throttle may then be opened and the engine started with the steam as dry as if it had been running continuously.

CHIMNEYS AND DRAUGHT.

Draught, in chimneys, is caused by the difference between the weight of the air outside and that inside the chimney. This difference in weight is produced by difference in heat.

Now, heated air has a strong tendency to rise above cool air and a very slight difference will cause an upward flow of the heated particles, and the hotter the air, the brisker the flow.

As these particles ascend it leaves a space which the cooler air eagerly hastens to fill; in the boiler furnace, the hot air pushing its way up the chimney, is replaced through the grate bars with cool, fresh air.

It is the mingling of this fresh air with the combustibles that produces heat, and the power of the draught is absolutely necessary to the reliable operation of the furnace.

An excess of draught can be corrected by the use of a damper or even by the closing of the ash pit doors, but no more unhappy position for an engineer can be imagined than a deficiency of draught.

This lack is produced by, 1st, too little area in the chimney flue; 2d, by too low a chimney; 3d, by obstructions to the flow of the gases; 4th, by the overtopping of the chimney by adjacent buildings, hills or tree tops. There are other causes of failure which practice develops; hence, the draught of a new chimney is very often an uncertain thing until every-day trial demonstrates its action.

The draught of steam boilers and other furnaces should be regulated below the grate and not in the chimney. The ash pit door should be capable of being closed air tight, and the damper in the chimney should be kept wide open at all times unless it is absolutely necessary to have the area of the chimney reduced in order to prevent the gases from escaping too fast to make steam.

When two flues enter a larger one at right angles to it, opposite each other, as is frequently the case where there is a large number of boilers in a battery, and the chimney is placed near the center of the battery, the main flue should always have a division plate in its center between the two entering flues to give direction to the incoming currents of gases, and prevent

CHIMNEYS AND DRAUGHT.

their "butting," as it may be termed. The same thing should always be done where two horizontal flues enter a chimney at the same height at opposite sides.

In stationary boilers the chimney area should be one-fifth greater than the combined area of all the tubes or flues.

For marine boilers the rule is to allow fourteen square inches of chimney area for each nominal horse power.

The draft of a chimney is usually measured in inches of water. The arrangement most commonly made use of for this purpose consists of a U-shaped glass tube connected by rubber tubing, iron pipe, or other arrangement, with some part of the chimney in such a way that the draft will produce a difference of level of water in the two legs of the bent glass tube.

The "Locomotive" suggests that *the unit for chimney construction* should be a flue 81 feet high above the level of the grates, having an area equal to the collective area of the tubes of all the boilers leading to it, the boilers being of the ordinary horizontal return tubular type, having about 1 square foot of heating surface to 45 square feet of heating surface.

Note the above conditions, and, in case of changing the above proportions, it should be observed that the draught power of chimneys is proportional to the square root of the height, so we may reduce its area below the collective area of the boiler tubes *in the same proportion that the square root of its height exceeds the square root of* 81.

For example, suppose we have to design a chimney for ten boilers, 66 in. in diameter, each having 72 tubes, 3½ in. in diameter, what would be its proportion.

The collective area of the 720 3½ in. tubes would be 6,017 square inches, and if the chimney is to be but 81 feet high, it should have this area, which would require a flue 6 ft. 5½ in. square.

But, suppose, for some reason, it is decided to have a chimney 150 feet in height, instead of 81 feet. The square root of 150 is 12¼; the square root of 81 is 9; and we reduce the area of the chimney by the following proportion: 12.25: 9=6,017: 4,420 square inches, which would be the proper area, and would call for a chimney 5 ft. 6 in. square, and similarly if any other height were decided upon.

PLUMBING.

The art of working in lead is older than the pyramids. For thousands of years hydraulics and plumbing as an occupation engaged the principal attention of engineers. King

David used lead pipe, so did Archimedes; the terraces and gardens of Babylon were supplied with water through leaden pipes. Steam

fitting, with galvanized pipe and an elaborate system of connections and devices is a new department of mechanism—almost

of the present generation—and at first sight would seem able soon to supercede lead piping of all kinds, but it is safe to say that nothing can ever take the place of lead, for this admirable metal can be made to answer where no other material can be worked; for instance, lead pipe can be made to conform to any angle or obstruction where no other system of piping will. Hence, plumbing as a useful and ornamental art will never go out of date, and engineers of every branch will do well to study its principles and methods so as to meet the ever-recurring and perplexing questions connected with sewerage, water supply, etc.

Every engineer should at least know how 1, *to join lead pipe*—to make a "wipe joint,"—as in a hundred emergencies this knowledge will be of worth. 2, how to make a temporary stopping of leaks; 3, how to bend pipe with sand or springs; 4, how to "back air pipes" from sinks; 5, how to use force pumps; 6, how to arrange the circulating pipes in hot-water boilers; 7, how to make solder; 8, how to repair valves, etc., etc.

PIPING AND DRAINAGE.

The three illustrations on page 298 are designed to represent traps set in lead pipe and show vividly the difference between this material and iron piping.

Lead is one of the elementary substances of which the world is formed; it ranks with gold, silver, tin, etc., in being an unmixed metal. It melts at about 617° Fahrenheit, and is, bulk for bulk, $11\frac{4}{10}$ heavier than water (gold being $17\frac{5}{10}$ heavier and wrought iron $7\frac{7}{10}$ heavier). The tenacity of lead is extremely low, a wire $\frac{1}{8}$th of an inch breaks with a weight of

Fig. 159.

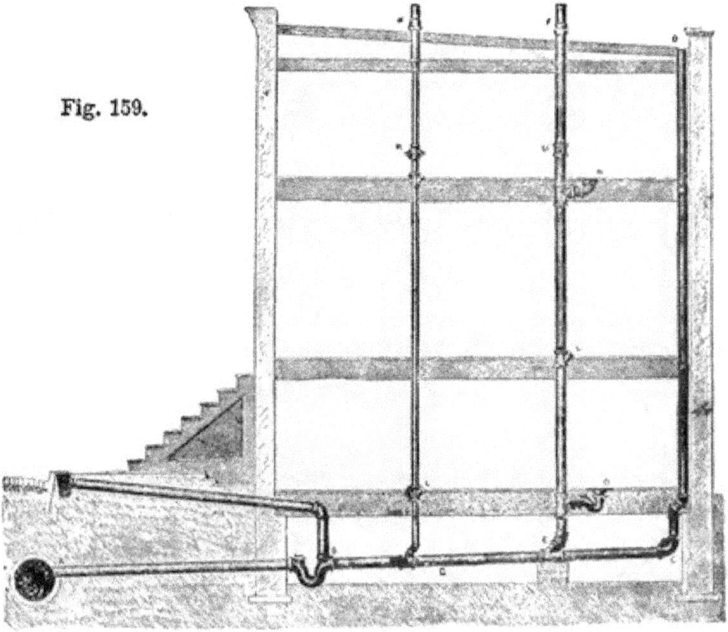

28 lbs.; in comparison, its tenacity is only one-twentieth that of iron; it is so soft that it may be scratched with the thumb nail. If a very strong heat is applied lead boils and evaporates; it transmits heat very slowly; of seven common metals it is the worst conductor, therefore it is good for hot water pipes. Mixed with a sufficient quantity of quicksilver it remains liquid.

An advantage to be found in the use of lead is its durability and comparative freedom from repairs. In London, soil and drain water pipes which have been fixed 300 to 500 years are as

PIPING AND DRAINAGE.

good now as the day they were first made—while iron pipe cannot be expected to last over 10 or 20 years or 30 at the utmost.

Fig. 159 represents the general system of house piping and drainage applicable also to shops, public buildings, etc. A exhibits the drain or sewer. A–C represents the sewer connection, so called with a running trap, B. "C" at the end of the lower pipe exhibits a soil pipe elbow, with hand hole for cleaning out closed by a screw plug. This drain should have a regular fall or inclination and this elbow provides for that. C–D shows the rain water leader (conductor.)

E and F is a soil pipe 3, 4, 5, or 6 inches in diameter. Note, pipes draining water closets are called "soil pipes"; those draining other fixtures "waste pipes." N and O represent water closet flanges; F and H are roof connections; L exhibits double and single **Y** branches to receive waste-pipes from baths, bowls, or sinks. The plumber makes this connection, always trapping the lead waste-pipe and then soldering it to a *brass* nipple.

LEAD PIPE JOINTS.

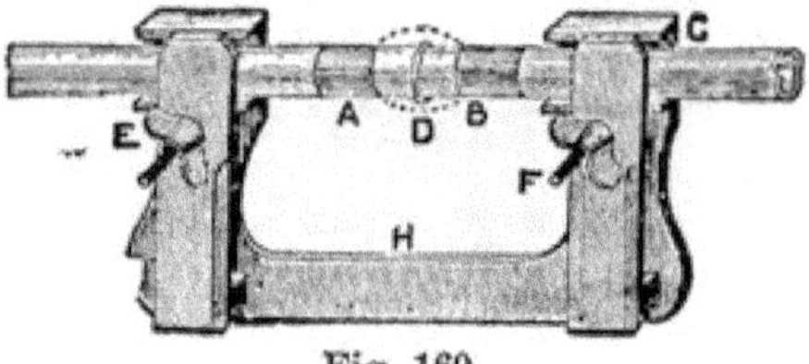

Fig. 160.

It has been remarked that after learning how to make "a wipe joint," everything is easy relating to the plumber's trade; hence, the importance of the following directions.

To learn the art, previous practice with short pieces of pipe is recommended. This trial piece can be clamped as shown in Fig. 160 and used over and over until practice has been had.

There are many names for the process of lead joint-making, such as the flow-joint, the ribbon joint, the blown joint, the astragal joint, etc., to express the different positions and uses

LEAD PIPE JOINTS.

for which they are needed, but in the main they are made as follows :

1. The lead pipe to be joined is sawn square off with the proper toothed saw—attention being paid to making the end absolutely true, across the pipe.

2. One end of the pipe to be joined is first opened by driving in a wooden wedge, shaped like a plumb-bob, called the "turn pin," Care should be exercised at this time not to split the end, ¼ inch opening is usually enough, which leaves the pipe as shown at D, Fig. 161. Now, clean the internal part of the joint all around the part required for soldering—this cleaning can be done with the plumber's shave hook or with a pocket knife. To complete this preparation "touch" the part with grease from a tallow candle.

3. Next is the preparation of the male part of the joint. This must be rasp-filed down to fit the enlarged opening. It is important to have a good fit throughout; hence, inside the enlarged opening must be also rasp-filed and the two surfaces to come nicely together before the solder is applied.

4. At this stage a paste called "plumber's soil" must be applied outside 3 inches from the end of each piece of pipe; this is shown by the line E F in Fig. 161, also at A B, Fig. 160; the line of the soiling should be very even and true in order to assure a workmanlike job and the soiling put on as before stated, 3 *to* 5 *inches beyond the solder line on each side.*

As the melting point of lead is 612 degrees or thereabouts, it is necessary to have solder melt at a lower temperature, and that made under the rule given will melt at 440 to 475 degrees.

No tool to a plumber is more important than the cloth used in joint making. To make it, take a piece of new mole skin or fustian, of moderate thickness, 12 inches long by 9 inches wide, fold it up one side 4 inches; then 4 inches again, and again 4 inches; then fold it in the middle, which will make your cloth 4x4½ inches, and of 6 thickness. After this is done, sew up the ragged ends to keep it from opening. Then pour a little hot tallow on one side and the cloth is ready for use. In Fig. 160-a is shown, H, a hand holding the cloth C in the process of "wiping the joint," which will now be described.

LEAD PIPE JOINTS.

First place a small piece of paper under the joint to catch the surplus solder D and begin soldering as follows: Take the felt F in the right hand and with it hold the ladle three parts full of solder. To see that it is not too hot hold your hand within 2 inches or so of the solder; if it quickly burns your hand it is too hot; if you can only just hold your hand this distance, use it; but if you cannot feel the heat, the solder is too cold.

When you begin to pour your solder upon the joint do it very lightly and not too much at a time in one place, but keep the ladle moving backward and forward, pouring from E to J, first on one side of the joint to the other and from end to end.

Pour also an inch or two up the soiling, as shown at E to make the pipe of proper temperature, *i. e.*, to the same heat as the solder. The further, in reason, the heat is run or taken along the pipe, the better the chance of making the joint.

Fig. 160-a.

Keep pouring and with the left hand hold the cloth C to catch the solder and also cause the same to tin the lower side of the pipe and to keep the solder from dropping down. This cloth, so important in joint making is elsewhere described. By the process of steady pouring the solder now becomes nice and soft and begins to feel shaped, firm and bulky.

When in this shape and in a semi-fluid condition quickly put the ladle down, and instantly with the left hand shape one side of the joint always beginning at the outsides, or at that part next the soiling; then take the cloth in the right hand and do the other side, *finishing on the top;* a light run of the cloth all round the joint will, if the solder has not set and you have been quick with your work, give the appearance of a turned joint. After a little practice the joint may be made without changing the cloth from one hand to the other.

The secret of joint making is getting the lead to the heat of the solder and in roughly shaping the solder, while in the semi-fluid state.

Good mechanical fitting is the result of two things—good judgment and a delicate sense of touch.

Repairing Pipes with Putty Joints.

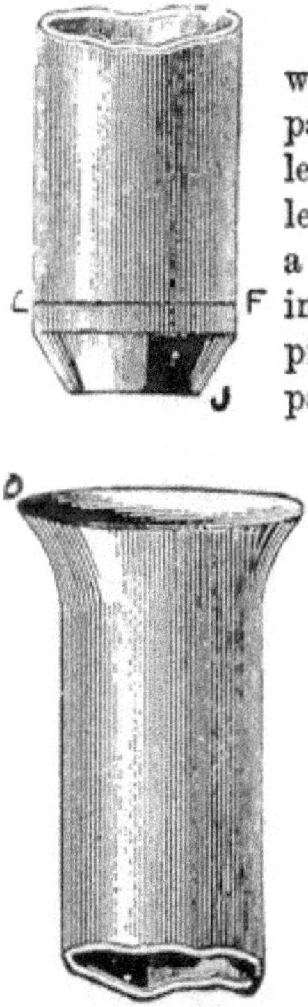

Fig. 161.

First get the pipe *thoroughly dried,* and with some quick drying gold size paint the part to be repaired; then get some white lead and stiffen it with some powdered red lead, so as to make it a hardish putty, place a thin layer of this, say $\frac{3}{8}$th inch to $\frac{1}{2}$ inch in thickness, over the bursted part of the pipe, and with some good strong calico, painted with the gold size, neatly wrap the red lead to the pipe, using 3 or 4 thicknesses of the painted calico; then with some twine begin at one end, laying the twine in several layers in rotation until it has, like the calico, several thicknesses.

If properly done this will be strong enough to withstand any ordinary pressure on the pipes and if more is required the putty can be made from dry red lead and gold size. In making all white and red lead joints, first, see that the parts are thoroughly dry; second, see that the parts are not dirty with rust, &c.; next, well paint the parts with good, stiff paint before putting the putty on to form the joint.

BENDING LEAD PIPE.

If any ordinary piece of light lead pipe 1½ inches in diameter is taken and pulled or bent sharply around it will crimple or crinkle at the throat; the larger and thinner the pipe the more it will become distorted.

There are many methods of making these bends in lead pipe, some with dummies, others with bolts, balls, etc., others cut the bends at the back, at the throat or the two sides of the bend.

For small pipes, such as ½ to 1 inch and extra heavy, they may be pulled round without trouble or danger, but for a little larger size SAND BENDING is largely practiced, as follows:

Take the length of pipe, say 5 feet, fill and well ram it with sand 2 feet up, then have ready a metal pot of very hot sand to fill the pipe 1 foot up, next fill the pipe up with more cold sand, ramming it as firmly as possible, stop the end and pull round the pipe, at the same time hammering quickly working the lead from the throat towards the back, which can be done if properly worked. N. B.—Care must be used not to reduce or enlarge the size of the bore at the bend.

BENDING WITH WATER.—It is a well-known fact that for such work, water is incompressible, but may be turned or twisted about for any shape provided it is enclosed in a solid case. To make the bend—the end of the pipe is stopped and a stop cock soldered into the other end; take the pipe at the end and pull it around, being careful that the water does not cool and shrink, and hammering quickly to take out the crinkle.

BENDING WITH BALLS.—This method is practiced with small pipe and also to take "dints" out in case of sand and water bending when a ball is sent through. Method: suppose the pipe to be 2 inches, then a ball is required $\frac{1}{16}$ in. less than the pipe, so that it will run through the pipe freely. Now pull the pipe round until it just begins to flatten, put the ball into the pipe and with some short pieces of wood, say 2 in. long by 1½ in. in diam., force the ball through the dented part of the pipe. The ball will run through all the easier if "touched" over with a candle end. Care must be used in forcing the ball back and forth not to drive it through the bend.

TABLE.—WEIGHT OF SHEET LEAD.

Inside Diameter........	⅜	½	⅝	¾	1	1¼	1½	1¾	2
AAA, weight per foot, lbs., oz.	2–8	3–0	3–8	4–12	6–0	...	....	....	....
AA, " "	1–8	2–0	2–12	3–12	4–12	6–0	8–0	8–8	9–0
A, " "	1–4	1–12	2–8	3–0	4–3	4–12	6–8	6–8	7–0
B, " "	1–4	1–4	2–0	2–4	3–4	3–12	5–0	5–0	6–0
C, " "	–10	1–0	1–8	1–12	2–8	3–0	4–4	4–0	4–12
D, " "	–7	–12	1–0	1–4	2–0	2–8	3–8	....	....
E, " "	....	–9	–12	1–0	1–10	2–0	3–0	....	....

Sheet lead is not the same weight, bulk for bulk, owing to difference in organic formation, but a cubic foot may be said to weigh 709 lbs.

A square foot 1″ thick, 59 "
" " " ⅛″ " 7½ "
" " " $\frac{1}{16}$″ " 6 "
" " " $\frac{1}{18}$″ " 5 "
" " " $\frac{1}{15}$″ " 4 "
" " " $\frac{1}{20}$″ " 3 "

Sheet lead is sometimes made as thin as writing paper.

PLUMBER'S SOLDER.

Rule for making.—Take 100 lbs. good old lead or lead cuttings, run it down thoroughly, stir it up and take off all dirt or dross: then take 50 lbs. pure tin, let this run down, and when nearly all is melted and is a little cooler throw in ½ lb. of black rosin, and well stir the lot up. Last bring up the heat to 600 degrees which may be known by the burning of a bit of newspaper put in the pot. The solder is now hot enough and should be well stirred and then run into moulds.

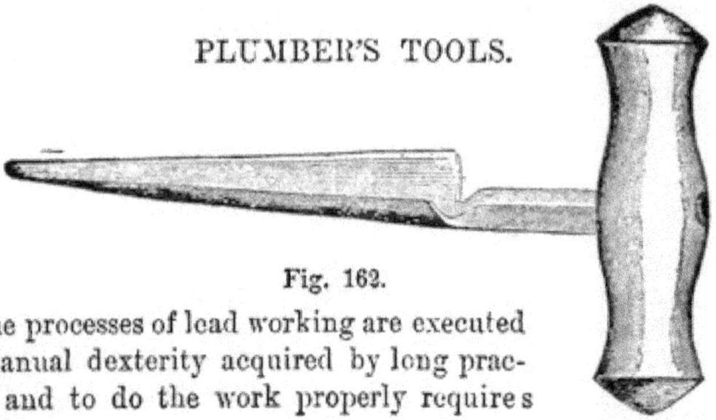

Fig. 162.

The processes of lead working are executed by manual dexterity acquired by long practice, and to do the work properly requires many special tools. Some of these are used in common with other departments of mechanics but are none the less necessary in lead working.

Fig. 163.

We present cuts of the principle tools used some of which are self-explaining and some are named with further desecription of particular use.

Fig. 164.

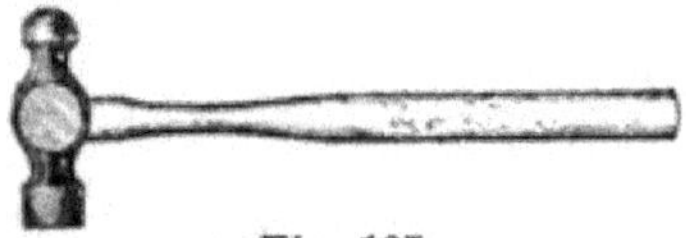

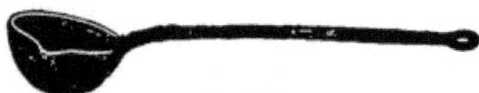

Fig. 165.

Fig. 162 represents one form of the plumber's tap borer or reamer, used for making and enlarging holes in pipe.

Fig. 166.

Fig. 163 represents plumber's snips.

Fig. 164 is the well-known and always useful ladle.

Fig. 165 is the round nose pene hammer, used in plumber's work to open the inside pipe before jointing.

Fig. 166 is the plumb bob. The same cut will also convey an idea of the wooden instrument used to force open the pipe before jointing, *i. e.*, "the turn pin."

PLUMBER'S TOOLS.

Fig. 167.

Fig. 167 represents "the round nose chisel."

Fig. 168.

Fig. 168 is the "wood chisel" used in cutting away wood work.

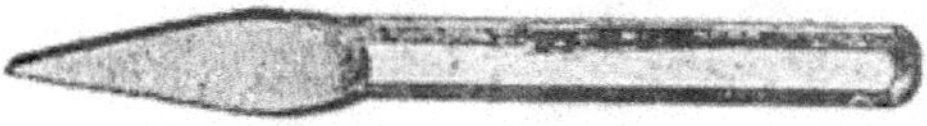

Fig. 169.

Fig. 169 is the half round nose chisel.

Fig. 170.

Fig. 170 is the well-known "cape chisel."

Fig. 171.

Fig. 171 is the equally well-known "cold chisel."

Fig. 172.

Fig. 172 is the "diamond nose chisel."

Fig. 173.

Fig. 173 shows a rivet set for small work connected with plumbing and sheet metal work.

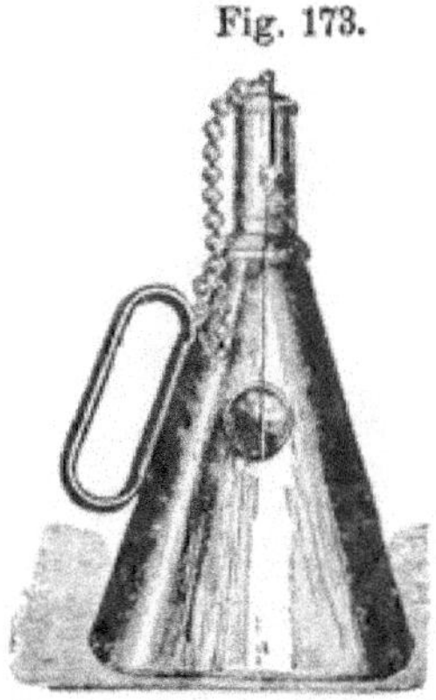

Fig. 174.

Fig. 174 exhibits the plumber's torch this is also used by engineers to explore interiors of boilers, chimney flues, and other dark places about the steam plant.

Fig. 175 is a compass saw.

Fig. 176 is a double edged plumber's saw.

Fig. 177 is a spirit level.

Fig. 178 is a looking-glass used in making underhand joints and in many useful ways about a steam plant.

PLUMBER'S TOOLS.

Fig. 175.

Fig. 176.

Fig. 177.

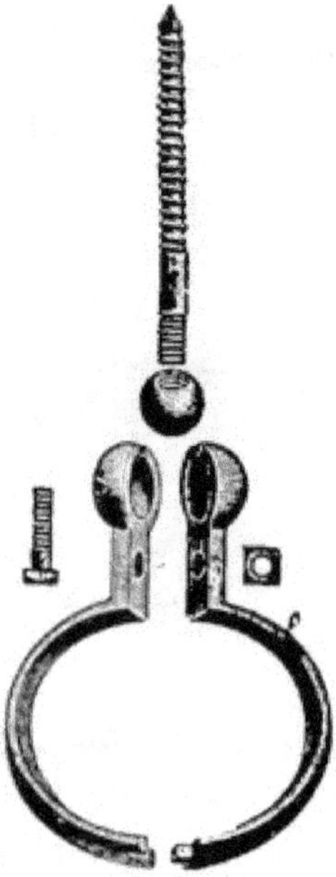

Fig. 183.

Fig. 178.

Fig. 179 is a caulking tool.

Fig. 180 is a gasket chisel.

Fig. 181 is a soldering tool known among plumbers as "a copper pointed bolt."

Fig. 182 is a copper-pointed bolt, flat.

Fig. 183 represents a hanger, for suspending iron and lead pipe; its excellence consists in enabling pipes to be raised or lowered after being hung without taking the hanger apart.

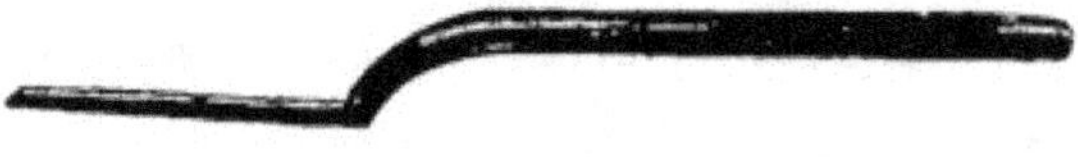

Fig. 179.

Fig. 180.

Fig. 181.

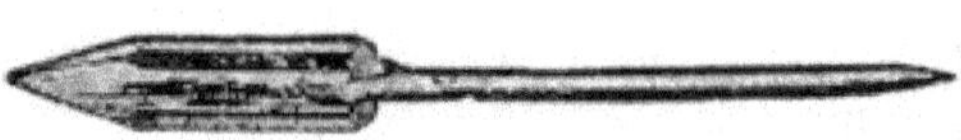

Fig. 182.

USEFUL TABLES OF WEIGHTS OF IRON AND COMPARISONS OF GAUGES.

Weight of a Superficial Foot of Plate and Sheet Iron

Plate Iron.		Sheet Iron.			
		United States Standard Guage. Adopted by Congress, to take effect July 1st, 1893.			
Thickness. INCHES.	Weight per square foot. POUNDS.	Number of Guage.	1000's of an inch.	Weight per square foot. OUNCES	Nearest fraction of an inch.
1/16 in.	2½	No. 1	.281	180 oz.	9/32 in.
1/8 "	5	" 2	.265	170 "	17/64 "
3/16 "	7½	" 3	.250	160 "	1/4 "
1/4 "	10	" 4	.234	150 "	15/64 "
5/16 "	12½	" 5	.218	140 "	7/32 "
3/8 "	15	" 6	.203	130 "	13/64 "
7/16 "	17½	" 7	.187	120 "	3/16 "
1/2 "	20	" 8	.171	110 "	11/64 "
9/16 "	22½	" 9	.156	100 "	5/32 "
5/8 "	25	" 10	.140	90 "	9/64 "
11/16 "	27½	" 11	.125	80 "	1/8 "
3/4 "	30	" 12	.109	70 "	7/64 "
13/16 "	32½	" 13	.093	60 "	3/32 "
7/8 "	35	" 14	.078	50 "	5/64 "
15/16 "	37½	" 15	.070	45 "	9/128 "
1 "	40	" 16	.062	40 "	1/16 "
		" 17	.056	36 "	9/160 "
		" 18	.050	32 "	1/20 "
		" 19	.043	28 "	7/160 "
		" 20	037	24 "	3/80 "
		" 21	034	22 "	11/320 "
		" 22	.031	20 "	1/32 "
		" 23	.028	18 "	9/320 "
		" 24	.025	16 "	1/40 "
		" 25	.021	14 "	7/320 "
		" 26	.018	12 "	3/160 "
		" 27	017	11 "	11/640 "
		" 28	.015	10 "	1/64 "
		" 29	014	9 "	9/640 "
		" 30	.012	8 "	1/80 "

USEFUL TABLES.

Weight of One Foot of Round Iron.

SIZE.	Weight pr. Foot.	SIZE.	Weight pr. Foot.	SIZE.	Weight pr. Foot.
	LBS.		LBS.		LBS.
1/8 in.	.041	1 7/16 in.	5.41	3 1/2 in.	32.07
3/16 ..	.092	1 1/2 ..	5.89	3 5/8 ..	34.40
1/4 ..	.164	1 9/16 ..	6.39	3 3/4 ..	36.82
5/16 ..	.256	1 5/8 ..	6.91	3 7/8 ..	39.31
3/8 ..	.368	1 11/16 ..	7.45	4 ..	41.89
7/16 ..	.501	1 3/4 ..	8.02	4 1/8 ..	44.55
1/2 ..	.654	1 13/16 ..	8.60	4 1/4 ..	47.29
9/16 ..	.828	1 7/8 ..	9.20	4 3/8 ..	50.11
5/8 ..	1.02	1 15/16 ..	9.83	4 1/2 ..	53.01
11/16 ..	1.24	2 ..	10.47	4 5/8 ..	56.00
3/4 ..	1.47	2 1/8 ..	11.82	4 3/4 ..	59.07
13/16 ..	1.73	2 1/4 ..	13.25	4 7/8 ..	62.22
7/8 ..	2.00	2 3/8 ..	14.77	5 ..	65.45
15/16 ..	2.30	2 1/2 ..	16.36	5 1/8 ..	68.76
1 ..	2.62	2 5/8 ..	18.04	5 1/4 ..	72.16
1 1/16 ..	2.95	2 3/4 ..	19.80	5 3/8 ..	75.64
1 1/8 ..	3.31	2 7/8 ..	21.64	5 1/2 ..	79.19
1 3/16 ..	3.69	3 ..	23.56	5 5/8 ..	82.83
1 1/4 ..	4.09	3 1/8 ..	25.57	5 3/4 ..	86.56
1 5/16 ..	4.51	3 1/4 ..	27.65	5 7/8 ..	90.36
1 3/8 ..	4.95	3 3/8 ..	29.82	6 ..	94.25

Weight of One Foot of Square Iron.

SIZE.	Weight pr. Foot	SIZE.	Weight pr. Foot.	SIZE.	Weight pr. Foot
	LBS.		LBS.		LBS.
1/8 in.	.052	1 7/16 in.	6.89	3 1/2 in.	40.83
3/16 ..	.117	1 1/2 ..	7.50	3 5/8 ..	43.80
1/4 ..	.208	1 9/16 ..	8.14	3 3/4 ..	46.88
5/16 ..	.326	1 5/8 ..	8.80	3 7/8 ..	50.05
3/8 ..	.469	1 11/16 ..	9.49	4 ..	53.33
7/16 ..	.638	1 3/4 ..	10.21	4 1/8 ..	56.72
1/2 ..	.833	1 13/16 ..	10.95	4 1/4 ..	60.21
9/16 ..	1.06	1 7/8 ..	11.72	4 3/8 ..	63.80
5/8 ..	1.30	1 15/16 ..	12.51	4 1/2 ..	67.50
11/16 ..	1.58	2 ..	13.33	4 5/8 ..	71.30
3/4 ..	1.87	2 1/8 ..	15.05	4 3/4 ..	75.21
13/16 .	2.20	2 1/4 ..	16.88	4 7/8 ..	79.22
7/8 ..	2.55	2 3/8 ..	18.80	5 ..	83.33
15/16 .	2.93	2 1/2 ..	20.83	5 1/8 ..	87.55
1 ..	3.33	2 5/8 ..	22.97	5 1/4 ..	91.88
1 1/16 ..	3.76	2 3/4 ..	25.21	5 3/8 ..	96.30
1 1/8 ..	4.22	2 7/8 ..	27.55	5 1/2 ..	100.80
1 3/16 ..	4.70	3 ..	30.00	5 5/8 ..	105.50
1 1/4 ..	5.21	3 1/8 ..	32.55	5 3/4 ..	110.20
1 5/16 ..	5.74	3 1/4 ..	35.21	5 7/8 ..	115.10
1 3/8 ..	6.30	3 3/8 ..	37.97	6 ..	120.00

USEFUL TABLES.

Weight per Running Foot of Cast Steel.

Size.	Lbs.	Size.	Lbs.	Size.	Lbs	Size.	Lbs.
¼ in. Sq.	.213	¼ in. Rd.	.167	1 × ¼	.852	½ in. Oct.	.745
½	.855	½	.669	1⅛ × ⅜	1.43	⅝	1.16
¾	1.91	¾	1.50	1¼ × ½	2.13	¾	1 67
1	3.40	1	2.67	1½ × ⅝	3.19	⅞	2 28
1¼	5.32	1¼	4 18	1¾ × ¾	4.46	1	2.98
1½	7.67	1½	6.02	2 × ½	3 40	1⅛	3.77
2	13.63	2 ..	10.71	.. × ⅝	4 25	1¼ .. .	4.65

Comparison of Principal Guages in use

Number.	United States Standard.		Stubbs' Birmingham.		Brown & Sharp.	
	1000's of an inch.	Pounds per square foot. IRON.	1000's of an inch.	Pounds per square foot. IRON.	1000's of an inch.	Pounds per square foot. IRON.
No. 1	.281	11.25	.300	12.04	.289	11.61
" 2	.265	10.62	.284	11.40	.257	10.34
" 3	.250	10.	.259	10.39	.229	9.21
" 4	.234	9.37	.238	9.55	.204	8.20
" 5	.218	8.75	.220	8.83	.181	7.30
" 6	.203	8.12	.203	8.15	.162	6.50
" 7	.187	7.50	.180	7 22	.144	5.79
" 8	.171	6.87	165	6 62	.128	5.16
" 9	.156	6.25	.148	5.94	.114	4 59
" 10	.140	5.62	.134	5.38	.102	4.09
" 11	.125	5.00	.120	4.82	.091	3.64
" 12	.109	4.37	.109	4.37	.080	3 24
" 13	.093	3.75	.095	3.81	.072	2.89
" 14	.078	3.12	.083	3 33	.064	2.57
" 15	.070	2 81	.072	2.89	.057	2.29
" 16	.062	2.50	.065	2.61	.050	2 04
" 17	.056	2.25	.058	2.33	.045	1.82
" 18	.050	2.00	.049	1.97	.040	1.62
" 19	.043	1.75	.042	1.69	.036	1 44
" 20	.037	1.50	.035	1 40	.032	1.28
" 21	.034	1.37	.032	1.28	.028	1.14
" 22	.031	1.25	.028	1.12	.025	1.02
" 23	.028	1.12	.025	1.00	.022	.90
" 24	.025	1.00	.022	.88	.020	.80
" 25	.021	.87	.020	.80	.018	.72
" 26	.018	.75	.018	.72	.016	.64
" 27	.017	.68	.016	.64	.014	.57
" 28	.015	.62	.014	.56	.012	.50
" 29	.014	.56	.013	.52	.011	.45
" 30	.012	.50	.012	.48	.010	.40

NOISELESS WATER HEATER.

This device is very effective for heating water in open or closed tanks by direct steam pressure without noise. The heater consists of an outward and upward discharging steam nozzle, covered by a shield which has numerous openings for the admission of water so that the discharge jet takes the form of an inverted cone, discharging upwards.

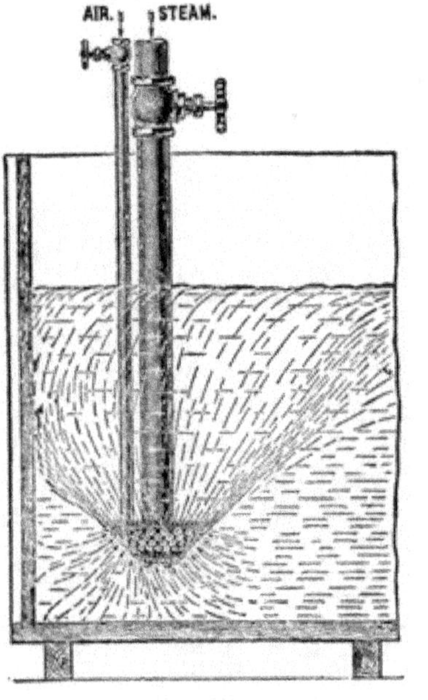

Fig. 184.

A small pipe admits air to the steam jet, and by mixing therewith prevents a collapse of the steam bubbles, and the noise, which is such a great objection to heating by direct steam in the old way. A valve or cock on the small air pipe regulates the opening as may appear most desirable.

Exhaust steam can by the same method be disposed of under water without noise.

ACCIDENTS AND EMERGENCIES.

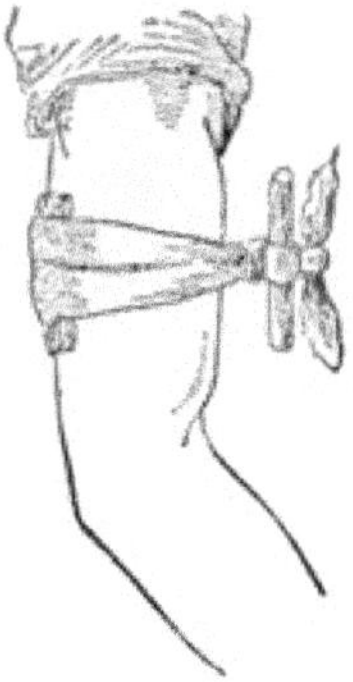

Fig. 184

Few subjects can more usefully employ the attention and study of engineers than the proper treatment and first remedies made necessary by the peculiar and distressing accidents to which persons are liable who are employed in or around a steam plant.

These and many other things of a like nature are likely to call for a cool head, a steady hand and some practical knowledge of what is to be done.

In the first moments of sudden disaster, of any kind, the thoroughly trained engineer is nearly always found, in the confusion incident to such a time, to be the one most competent to advise and direct the efforts made to avert the danger to life limb or property, and to remedy the worst after effects.

To fulfil this responsibility is worth much previous preparation, so that the best things under the circumstances may be done quickly and efficiently. To this end the following advice is given relating to the most common accidents which are likely to happen, in spite of the utmost exercise of care and prudence.

Burns and Scalds.—*Burns* are produced by heated solids or by flames of some combustible substance; *scalds* are produced by steam or a heated liquid. The severity of the accident depends mainly, 1, on the intensity of the heat of the burning body, together with, 2, the extent of surface, and, 3, the vitality of the parts involved in the injury, thus: a person may have a finger burned off with less danger to life than an extensive scald of his back.

The immediate effect of scalds is generally less violent than that of burns; fluids not being capable of acquiring so high a temperature as some solids, but flowing about with great facility their effects become most serious by extending to a large surface of the body. A burn which instantly destroys the part

ACCIDENTS AND EMERGENCIES.

which it touches may be free from dangerous complications, if the injured part is confined within a small compass; this is owing to the peculiar formation of the skin.

The skin is made up of two layers; the outer one has neither blood vessels nor nerves, and is called the scarf-skin or cuticle; the lower layer is called the true skin, or cutis. The latter is richly supplied with nerves and blood vessels, and is so highly sensitive we could not endure life unless protected by the cuticle. The skin, while soft and thin, is yet strong enough to enable us to come in contact with objects without pain or inconvenience.

The extent of the surface involved, the depth of the injury, the vitality and sensibility of the parts affected must all be duly weighed in estimating the severity and danger of an accident in any given case.

In severe cases of burns or scalds the clothes should be removed *with the greatest care*—they should be carefully cut, at the seams, and not pulled off.

In scalding by boiling water or steam, cold water should be plentifully poured over the person and clothes, and the patient then be carried to a warm room, laid on the floor or a table but not put to bed, as there it becomes difficult to attend further to the injuries.

The secret of the treatment is to avoid chafing, *and to keep out the air.* Save the skin unbroken, if possible, taking care not to break the blisters; after removal of the clothing an application, to the injured surface, of a mixture of *soot and lard,* is, according to practical experience, an excellent and efficient remedy. The two or three following methods of treatment also are recommended according to convenience in obtaining the remedies.

Take ice well crushed or scraped, as dry as possible, then mix it with fresh lard until a broken paste is formed; the mass should be put in a thin cambric bag, laid upon the burn or scald and replaced as required. So long as the ice and lard are melting there is no pain from the burn, return of pain calls for a repetition of the remedy.

BURNS AND HEAT STROKES.

The free use of soft soap upon a fresh burn will remove the fire from the flesh in a very little time, in ¼ to ½ an hour. If the burn be severe, *after relief from the burn,* use linseed oil and then sift upon it wheat flour. When this is dried repeat the oil and flour until a complete covering is formed. Let this dry until it falls off, and a new skin will be formed without a scar.

In burns with lime, soap lye, or *any caustic alkali,* wash abundantly with water (do not rub), and then with weak vinegar or water containing a little sulphuric acid; finally apply oil, paste or mixture as in ordinary burns.

It would be well to always keep ready mixed an ointment for burns; in fact a previous readiness for an accident robs it of half its ill effects.

GLUE BURN MIXTURE.

A method in use in the N. Y. City Hospital known as the "glue burn mixture" is composed as follows: "7½ Troy oz. white glue, 16 fluid oz. water, 1 fluid oz. glycerine, 2 fluid drachms carbolic acid. Soak the glue in the water until it is soft, then heat on a water bath until melted; add the glycerine and carbolic acid and continue heating until, in the intervals of stirring, a glossy strong skin begins to form over the surface. Pour the mass into small jars, cover with parafine papers and tin foil before the lid of the jar is put on and afterwards protect by paper pasted round the edge of the lid. In this manner the mixture may be preserved indefinitely.

When wanted for use, heat in a water bath and apply with a flat brush over the burned part."

Insensibility from Smoke.—To recover a person from this dash cold water in the face, or cold and hot water alternately. Should this fail turn the patient on his face with the arms folded under his forehead; apply pressure along the back and ribs and turn the body gradually on the side; then again slowly on the face, repeating the pressure on the back: continue the alternate rolling movements about sixteen times a minute until breathing is restored. A warm bath will complete the cure.

TREATMENT OF CUTS AND WOUNDS.

Heat-stroke or Sun-stroke.—The worst cases occur where the sun's rays never penetrate and are caused by the extreme heat of close and confined rooms, overheated workshops, boiler rooms, etc. The symptoms are, 1, a sudden loss of consciousness; 2, heavy breathing; 3, great heat of the skin; and 4, a marked absence of sweat.

Treatment.—The main thing is to lower the temperature. To do this, strip off the clothing, apply chopped ice wrapped in flannel to the head; rub ice over the chest, and place pieces under the armpits and at the side. If no ice can be had use sheets or cloths wet with cold water, or the body can be stripped and sprinkled with cold water from a common watering pot.

Cuts and Wounds.—In these the chief points to be attended to are: 1, arrest the bleeding; 2, remove from the wound all foreign bodies as soon as possible; 3, bring the wounded parts opposite to each other and keep them so; this is best done by means of strips of adhesive plaster, first applied to one side of the wound and then secured to the other; these strips should not be too broad, and space must be left between the strips to allow any matter to escape. Wounds too extensive to be held together by plaster must be stitched by a surgeon, who should always be sent for in all severe cases.

For washing a wound, to every pint of water add 2½ teaspoonfuls of carbolic acid and 2 tablespoonfuls of glycerine—if these are not obtainable, add 4 tablespoonsful of borax to the pint of water—wash the wound, close it, and apply a compress of a folded square of cotton or linen; wet it in the solution used for washing the wound and bandage down quickly and firmly. If the bleeding is profuse, a sponge dipped in very hot water and wrung out in a cloth should be applied as quickly as possible—if this is not to be had, use ice or cloth wrung out in ice water.

Wounds heal in two ways. 1, rapidly by primary union, without suppuration, and leaving only a very fine scar. 2, slowly by suppuration and the formation of granulations and leaving a large red scar.

ACCIDENTS AND EMERGENCIES.

Bleeding.—This is of three kinds: 1, from the arteries which lead from the heart; 2, that which comes from the veins, which take the blood back to the heart; 3, that from the small veins which carry the blood to the surface of the body. In the first, the blood is bright scarlet and escapes as though it was being pumped. In the second, the blood is dark red and flows away in an uninterrupted stream. In the third, the blood oozes out. In some wounds all three kinds of bleeding occur at the same time.

The simplest and best remedy to stop the bleeding is to apply direct pressure on the external wound by the fingers. Should the wound be long and gaping, a compress of some soft material large enough to fill the cavity may be pressed into it; but this should always be avoided, if possible, as it prevents the natural closing of the wound.

Pressure with the hands will not suffice to restrain bleeding in severe cases for a great length of time and recourse must be had to a ligature; this can best be made with a pocket handkerchief or other article of apparel, long enough and strong enough to bind the limb. Fold the article neck-tie fashion, then place a smooth stone, or anything serving for a firm pad, on the artery, tie the handkerchief loosely, insert any available stick in the loop and proceed to twist it, as if wringing a towel, until just tight enough to stop the flow. Examine the wound from time to time, lessen the compression if it becomes very cold or purple, or tighten up the handkerchief if it commences bleeding.

Some knowledge of anatomy is necessary to guide the operator where to press. Bleeding from the head and upper neck requires pressure to be placed on the large artery which passes up beside the windpipe and just above the collar bone. The artery supplying the arm and hand runs down the inside of the upper arm, almost in line with the coat seam, and should be pressed as shown in Fig. 184. The artery feeding the leg and foot can be felt in the crease of the groin, just where the flesh of the thigh seems to meet the flesh of the abdomen and this is the best place to apply the ligature. In arterial bleeding the pressure must be put between the heart and the wound, while in *venous* bleeding it must be beyond the wound to stop the flow as it goes towards the heart.

In any case of bleeding, the person may become weak and faint; unless the blood is flowing actively this is not a serious sign, and the quiet condition of the faint often assists nature in staying the bleeding, by allowing the blood to clot and so block up any wound in a blood vessel. Unless the faint is prolonged or the patient is losing much blood, it is better not to hasten to relieve the faint condition; when in this state anything like excitement should be avoided, external warmth should be applied, the person covered with blankets, and bottles of hot water or hot bricks applied to the feet and arm-pits.

Frost-bite.—No warm air, warm water, or fire should be allowed near the frozen parts until the natural temperature is nearly restored; rub the affected parts gently with snow or snow water in a cold room; the circulation should be restored very slowly; and great care must be taken in the after treatment.

Broken Bones.—The treatment consists of, 1, carefully removing or cutting away, if more convenient, any of the clothes which are compressing or hurting the injured parts; 2, very gently replacing the bones in their natural position and shape, as nearly as possible, and putting the part in a position which gives most ease to the patient; 3, applying some temporary splint or appliance, which will keep the broken bones from moving about and tearing the flesh; for this purpose, pieces of wood, pasteboard, straw, or firmly folded cloth may be used, taking care to pad the splints with some soft material and not to apply them too tightly, while the splints may be tied by loops of rope, string, or strips of cloth; 4, conveying the patient home or to a hospital.

To get at a broken limb, or rib, the clothing must be removed, and it is essential that this be done without injury to the patient; the simplest plan is to rip up the seams of such garments as are in the way. Boots must be cut off. It is not imperatively necessary to do anything to a broken limb before the arrival of a doctor except to keep it perfectly at rest.

ACCIDENTS AND EMERGENCIES.

Poultices.—These outward applications are useful to relieve sudden cramps and pains due to severe injuries, sprains and colds. The secret of applying a mustard is to apply it hot and keep it so by frequent changes—if it gets cold and clammy it will do more harm than good. Poultices to be of any service and hold its heat should be from one-half to one inch thick. To make it, take flaxseed, oatmeal, rye meal, bread, or ground slippery elm; stir the meal slowly into a bowl of boiling water, until a thin and smooth dough is formed. To apply it, take a piece of old linen of the right size, fold it in the middle; spread the dough evenly on one half of the cloth and cover it with the other.

To make a "mustard paste" as it is called, mix one or two tablespoonfuls of mustard and the same of fine flour, with enough water to make the mixture an even paste; spread it neatly with a table knife on a piece of old linen, or even cotton cloth. Cover the face of the paste with a piece of thin muslin.

How to Carry an Injured Person.—In case of an injury where walking is impossible, and lying down is not absolutely necessary, the injured person may be seated in a chair, and carried; or he may sit upon a board, the ends of which are carried by two men, around whose necks he should place his arms so as to steady himself.

Where an injured person can walk he will get much help by putting his arms over the shoulders and round the necks of two others.

A seat may be made with four hands and the person may be thus carried and steadied by clasping his arms around the necks of his bearers.

If only one person is available and the patient can stand up, let him place one arm round the neck of the bearer, bringing his hand on and in front of the opposite shoulder of the bearer. The bearer then places his arm behind the back of the patient and grasps his opposite hip, at the same time catching firmly hold of the hand of the patient resting on his shoulder, with his other hand; then by putting his hip behind the near hip of the patient, much support is given, and if necessary, the bearer can lift him off the ground and as it were, carry him along.

ACCIDENTS AND EMERGENCIES.

To carry an injured person by a stretcher (which can be made of a door, shutter, or settee—with blankets or shawls or coats for pillows) three persons are necessary. In lifting the patient on the stretcher *it should be laid with its foot to his head*, so that both are in the same straight line; then one or two persons should stand on each side of him, and raise him from the ground, slip him on the stretcher; this to avoid the necessity of any one stepping over the stretcher, and the liability of stumbling. If a limb is crushed or broken, it may be laid upon a pillow with bandages tied around the whole (*i. e.*, pillow and limb) to keep it from slipping about. In carrying the stretcher the bearers should "break step" with short paces; hurrying and jolting should be avoided and the stretcher should be carried so that the patient may be in plain sight of the bearers.

PERSONAL.

The fireman, so called, in steam service of any description, should and does on the average receive double the compensation of a man who has only his labor to bargain for.

In addition, he exercises his skillful vocation in sheltered places and is almost the last of the employees of a plant to be "laid off" and is certainly the first to be called on again after stoppage.

Still further, the fireman has an almost equal opportunity, with the best shop trained machinist, for advancement to the position of engineer in charge of the most extensive steam plants.

Now! this increased pay over ordinary labor and other numerous advantages accruing from the position, demand a generous return, and in ending this work, the author suggests these "points" for observance to the aspiring student, whether engineer, fireman, or machinist, namely—that sobriety should be held one of the first elements of strict observance; an unresting tidiness of person and premises; dignity of conduct, as being owed to the rising profession of steam engineering; and lastly, an unswerving fidelity of trust, which may include honesty, truthfulness and courage.

INDEX.

PERSONAL.

> For wise men the joys of reading are life's crowning pleasures. Books are our universities, where souls are the professors. Books are the looms that weave rapidly man's inner garments. Books are the levelers—not by lowering the great, but by lifting up the small.
>
> *Newell Dwight Hillis in "A Man's Value to Society."*

Professor Hawkins has long been most favorably known as a practical and helpful writer upon Steam and Electrical Engineering, and now, as the publishers of his works, we take pleasure in handing you this little "booklet" describing the five books which comprise his scientific works issued up to the present time all of which are thoroughly up-to-date.

We recommend them :

1. To all who come in contact in any way with Steam or Electricity in any of its very numerous industrial departments.
2. To anyone looking for advancement.
3. As books of standard reference on the subjects treated upon.

Soliciting your kind patronage we are respectfully yours,

Theo Audel & Co

N. Y. City, 63 5th Ave. Publishers

CONTENTS.

The Dynamo; Conductors and Non-Conductors; Symbols, abbreviations and definitions relating to electricity; Parts of the Dynamo; The Motor; The Care and Management of the Dynamo and Motor.

Electric Lighting; Wiring; The rules and requirements of the National Board of Underwriters in full; Electrical Measurements.

The Electric Railway; Line Work; Instruction and Cautions for Linemen and the Dynamo Room; Storage Batteries; Care and Management of the Street Car Motor; Electro Plating.

The Telephone and Telegraph; The Electric Elevator; Accidents and Emergencies, etc., etc.

The full one-third part of the whole work has been devoted to the explanation and illustrations of the dynamo, and particular directions relating to its care and management;—all the directions are given in the simplest and most kindly way to assist rather than confuse the learner. The names of the various parts of the machine are also given with pictorial illustrations of the same.

In the Catechism no less than 25 full page illustrations have been given of the various dynamo machines made in different parts of the country, and an equal number of part page illustrations.

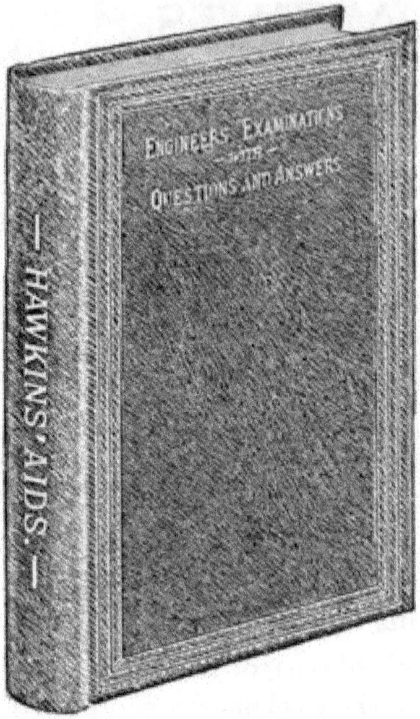

Engineers Examinations with Questions and Answers. Price $2.00.

The volume is bound (being designed for constant and ready reference) in substantial red leather with titles and edges in gold; it consists of between 200 and 300 pages, printed on heavy, fine surface paper; size 5″x7½″; weight 1½ lbs.

This book is a most important aid to all engineers, and is undoubtedly the most helpful ever issued relating to a safe and sure preparation for examination.

It presents in a condensed form the most approved practice in the care and management of Steam Boilers, Engines, Pumps, Electrical and Refrigerating Machines.

The following is a complete list of its contents:

> TESTIMONIAL.—FRED. D. STONE. "Inclosed find $2.00 for Hawkins' New Catechism of Electricity. I have HAWKINS' AIDS TO ENGINEERS EXAMINATIONS. It is the best on the subject. I owe my success in securing a license to it.

Contents.

1. This book embraces information not else where obtainable.

2. It tells exactly what an engineer will have to go through in getting a license, with much kindly and helpful advice to the applicant for a license.

3. It contains the annual report of the superintendents of "Steam Boiler Inspection and Certification of Engineers' for the cities of New York and Brooklyn.

4. It contains various rules, regulations and laws of cities for the examination of boilers and the licensing of engineers.

5. It contains the laws and regulations of the U. S. for the examination and grading of all marine engineers.

6. It gives a short chapter on the "Key to Success" in obtaining knowledge necessary for advancement in engineering. This is very important.

7. The book gives the underlying principles of steam-engineering in plain language, with sample questions and answers likely to be asked by the examiner.

8. It gives a few plain rules of arithmetic with examples of how to work the problems relating to the safety valve, strength of boilers and horse power of the steam engine and steam boiler.

9. The main subjects treated, upon which are given detailed information with questions and answers are as follows:—The Steam Boiler, Boiler Braces, Incrustation and Scale, Firing of Steam Boilers, Water Circulation in Boilers, Construction and Strength of Boilers, The Steam Engine, Engine and Boiler Fittings, Pumps, The Injector, Electricity and Electric Machines, Steam Heating, Refrigeration, Valve Setting, etc., etc.

CONTENTS.

Fire Irons; The firing of Steam-boilers; A Chapter of "Don'ts" relating to firing; Steam-boiler; the History and Growth of the Steam-boiler; Upright Boiler; the Marine Boiler; The Water-tube Boiler; The Locomotive Boiler; The Horizontal Tubular Boiler; Parts of the Boiler; Specification for 125 H. P. Boilers; Construction, Riveting and Bracing of Boilers, with many illustrations and tables for calculating the strength of same; Boiler Makers' Tools; Boiler Fixtures and belongings; Plumbing; Piping; Accidents and Emergencies; Index with nearly 1,000 References, etc., etc.

No Engineer, Fireman or Steam User can afford to be without this valuable book, as it contains the pith and vital "points" of economical and safe steam production.

The plan followed in this work is the same as that so generally approved in "Calculations"; it proceeds from the most simple rules and maxims to the highest problems; it is both a book of instruction and reference. The carefully-prepared Index contains nearly one thousand references, thus making it almost a dictionary of terms.

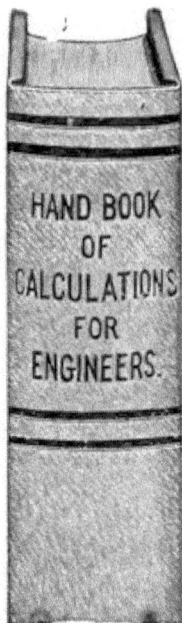

CONTENTS.

TESTIMONIALS.

"I am pleased with the work; it is of value to me. I have charge of a Harris-Corliss engine doing 680 H. P. at Slater's Cotton Mills."—Cyrus Bucklin, Pawtucket, R. I. * * "I think it is the best I ever saw, and I thank the day I saw it advertised."—Jno. C. Robinson, Adams, Mass. * * "The Hand Book is worth its weight in dollars to any engineer with common sense."—Jas. C. Temple, Eng., Springfield, Ill.

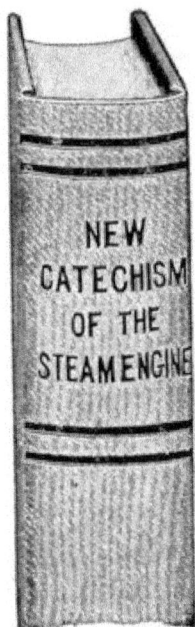

CONTENTS.

The subject matter of the New Catechism of the Steam Engine is not arranged in chapters, but according to the more natural order best designed to explain at greater or less length the different themes discussed. The following are the leading divisions of the 480 pages of the book.

Bindings.

Should a finer binding, than cloth, be desired of the "Calculations," "Boiler-Room," and "Steam Engine," we can supply them in elegant, full red leather bindings, gold titles and gilt edges—with the other two books, which are always bound in "leather and gold," as described; this binding makes the handsomest set of engineers' books on the market—as well as the most useful.

The extra cost is one dollar per volume (on the three books above named) over the cost of the regular cloth binding which is also very choice. The leather edition is also sold on part payments, separately or together.

A Few Selected Commendations.

There are no better judges of the true worth of these books than practical engineers. The following are examples of very many hundreds of recommendations sent without solicitation :—

Jos. S. Hall, Sebo, Mont., "The books are all right and should be used by every engineer"; *H. Chambers*, Atkinson, Neb., "The value of your books is beyond expression, may your good work continue"; *C. L. Wain*, Chief Engineer, Kamloops, B. C., "They are books that should be in the hands of every engineer"; *Adolph Hahn*, St. Louis, Mo., "I must say I am well pleased with the books, they are just what I was looking for"; *T. J. McCartney*, Anaconda, Mont., "Accept my thanks for the condition and promptness with which you delivered the books to me"; *T. A. Secord*, Unionville, Mo., "I take pleasure in saying that Hawkins' complete works are a very valuable addition to any engineer's library"; *William J. Lee*, Machinist, U. S. Steamer *Indiana*, off Havana, Cuba, April 26, '98, "The war is on us and so you will know the cause of any delay in payments, the men are very much pleased with the books, they say your books were just what they wanted. Two more sets are wanted."

Easy Payments.

The Hawkins' works are sold on easy part payments. The set (5 books) are sent upon the following terms: Upon the receipt of $2.00 and the promise to pay one dollar per month, the books will be forwarded, charges prepaid, to the purchaser in any part of the world; should one, two, or three books be desired, the payment in that case may be one dollar with order and the promise to pay the amount remaining due, one dollar per month.

The books are delivered free upon receipt of the first installment payment.

There is only one condition attached to this liberal offer. We require a reference as to the character and reliability of the intending purchaser, but in place of this a statement of where employed, how long and in what position, will be sufficient.

As soon as a payment is received a letter is written in acknowledgment, and in this letter a "safety envelope" will also be sent for the next installment due.

A printed order-blank with agreement, is furnished upon request, but this is not necessary. A simple order to send the books (with reference or statement) with the first payment will receive prompt attention and shipment.

The time to order these books is *now*, as the very pith of this method of payment is to avoid the waste of valuable time in studying them. This always has to be done *in advance* of "getting a raise" or securing a new position.

A long experience has shown the satisfactory workings of this plan of payments. It has increased the amount of the sales and put the advantages to be derived from the use of the books within the easy reach of hundreds of worthy persons. It is easily true that one thousand men hold advanced situations who owe them to these useful books. If one hundred dollars per year is the average increased pay for each it will show an added income of one hundred thousand dollars annually for our patrons. More than twenty-five thousand of the books have been sold to the date of the issue of this catalogue, each one of which has been a lifter (or lever) for the attentive student of its contents.

www.ingramcontent.com/pod-product-compliance
Lightning Source LLC
Chambersburg PA
CBHW021114270326
41929CB00009B/868
* 9 7 8 3 7 4 4 7 5 8 4 2 0 *